Matter, Information and Life

The principles are immutable, their interpretation what we want.

Juan D. Arango R.

Editorial Assistance:

Catalina Arango

Graphical Assistance:

Carolina Arango

ISBN: 978-0-9831683-1-7

Dedication

To my wife, Clara, who has been like the wind for the kite.

To my daughters, Carolina and Catalina, with whom I have learned so much.

To my family and friends, who have supported me in this project.

Content

Preface

"Never before has it become as necessary as now, that the new man's social stream is developing, to perform systematic and objective scientific research on human nature. Never before, too, has it been so desirable that data obtained by science are put to the service and benefit of as many people as possible to help alleviate their grief." Emilio Mira y López.

The purpose of this book is to bring the reader to think more comprehensively. Making use of information in a comprehensive manner starts with the understanding that neither matter nor information individually let you live, you require both. Matter defines your functional requirements; information is used to understand your needs (requirements) but also creates your desires (objectives). At the same time, this book is talking about the outstanding ability of human beings over other living beings, advanced abstraction ability, which is not yet adequately addressed; we need to understand that as humankind, we are a living being and every one of us, a part of it.

Thomas Kuhn introduced the concept of a paradigm, a filter in our mind which directs our perception of nature. Katherine Benziger has extended the idea of this filter to four different ways to process facts. The point of view presented in this book requires us to look at nature more candidly, not as kings of nature, but as a part of it. We are matter that has the same properties as any other matter, referring to the atoms that make us up. But a specific process, the informational ability, makes us living beings. Of course, it is not easy to understand, when someone come to us with new ideas, because each person has his own model, one that has taken him to where he is, that is familiar and frequently used in all his actions. Whichever your way of seeing the world, there is always room for improvement –this at least is what progressive people think– that the world of tomorrow can be better in every way.

I intend to achieve my goal: to show that freedom and justice depend on people, and there will be more peace and happiness in everyone if we take responsibility. I will try my best to fit the writer's role. Although I would like an impeccable book, I am far from a master of the written word, so, I just want to express my arguments accurately and easily reach the wise, while not making it hard work for others. Since perfection does not exist, I cannot say I want freedom and justice for all people, because as you know, freedom and justice are personal concepts. Each human being has its own idea of them. Now, if I make you understand that we are all responsible for our own lives through our decisions, and there will be more peace and more happiness if we get to act with more justice toward ourselves and others, my work will have taken course. Nobody can take away the freedom to make our own decisions. Only lack of faith and courage allow that we submit to unfair treatment from others; it is not easy, but it is in our hands to live better.

I content myself with helping to address change toward human justice. As is well known, perfect justice is from another world, the informational world where everything can be perfect, and this is the human world, built on the nature of matter, I say it with all certainty. Most humans are making change in one direction: irrational, instinctive; we are like most other animals, who live from emotions, involving the facts. More awareness is needed, not in the rational way, but in the philosophical way. There are sophists who try to confuse the common good to being equal; we need a better life for all, not an equal life for everyone. I do not want to talk about Communism, which is so "altruistic", because it annoys me as much as Capitalism, which is so "selfish". Beauty is in both justice and joy; in justice, which makes human beings of character, as in the joy of a child, which fills us with satisfaction; as well as in parties where people dance, sing and make friends; there is also beauty in the satisfaction of accomplishment, when we process natural resources and build bridges or roads to improve communication without destroying the environment, that is, when we transform matter.

Matter is the basis of everything we see and the basis of our substance. This is denied by some who talk about something before the matter. How is it possible? Everything is possible in the world of information; we are the Kings of the informational world. When have we seen someone with less informational capacity, which we call animals, laugh at a joke? Likewise, when have we seen an altar in the territory shared by a herd of chimpanzees? Those who deny matter as the basis of all that exists, usually demerit the informational ability of other living beings. But all living beings have informational ability or they would be non-living beings! Then the difference is in the quality, quantity and the goal setting of the informational ability. Measuring the quantity is an informational issue that we have learned to manage with the creation of numbers, not a matter of whether we can think or live along instincts. Our vanity as humans leads us to say we are the favorite creatures of nature. The reality is very far away from that idea. All strength, every tool, comes with its own weaknesses. Having more developed informational ability has made us the Kings of nature, but it has created its own problems: We carry the burden of duty, kill like no other animal and as widely as we can, and after that we justify or create grounds for having done so, and we get off the hook without the least shame.

Even today, the 21st century, the properties of matter are still a mystery. What can we expect then of the properties of living beings and their ability to handle information! We have no clear structure of matter, for the true basis of all beings. How are we going to have a clear understanding of material structures which preceded the formation of living beings! How we are going to be able to clearly understand the dynamics of the brain or other organs of the body!

Information is the result of the informational ability, the process we call life (Arango, 2010). Information is a unique and particular process. It is unique for all living beings and particular to each of them; there are no two living beings who have lived the same experience. You will say that you have lived the same experience of your company at a party. But you have not, you will look slightly different to your counterpart perspective, you are in a different place. In addition, each has his own particular truth, you have yours and your partner has his own. Both truths are different, complementary in some respects, parallel in others, but always particular. The information age has been upon the Earth since the first string of life, and its quality is the same abstract, playful, and with a particular meaning for who possesses it. The information creates a philosophy of life and a way to run our lives, a way of life.

The scientific method taught us to isolate the world and has been and will continue to be useful. However it is time to add a systemic approach to our lives and understand parts interact to create new properties, which are not evident when decomposing the system. In addition, these emergent properties do not comply with logical laws. Many are additive and there are options between black and white, colorful[1] choices, which represents many more options than the

---

[1] It was mentioned by my daughter, Catalina, during the process of discussing the book. This has several interesting aspects because one usually refers to grey as the combinations arising between black and white. But by decomposition of light, white is the composition of all colors, having more options to colors than in grayscale.

extremes, is or isn't, that some people see as: you are with me or against me. We must live, with multidisciplinary approaches (several view points at the same time), to achieve the strongest idea (different interpretations to such analyses); in other words, the concept of information and the concept of life cannot be understood without understanding the interactions between the parts that create emergent properties, a combined effect. The challenge is great, but no more than other challenges we have already overcome. We should understand that the movements of elemental particles define processes, and systems are defined as an instant state of these processes. Today, we do not understand the details in many natural processes including fire and life. Why do we consider life to be supernatural? This just shows us that we must improve our awareness and become conscious of the difference among matter, actions over it and the result of them.

Mankind has already passed the limit of the physical need. I say this in the sense that the mechanization of work today allows the release of labor, or should I say release of brains; so applying more intelligence in more productive ways and working in more areas of knowledge with more human sense targets and in a more systematical way, we can reach more distant sites into above, like other galaxies, or into down, like discovering the structure of matter. In addition, now that mechanization and production tools let us use the mind beyond what we have done before, it is time to speak more clearly with our awareness, define who's who and where the limits of freedom and justice are. We should have the freedom to choose a practical or idealistic venture according to our own abilities and not a pessimistic or optimistic one with the awareness of another, even though that person is someone close to us. Justice is not the law, justice must be seen in the streets, we are talking about real justice, where each one has what he deserves, not equality. Everyone as part of a community must have basic means –knowing that is consciousness– and we can all take risks between the practical and the idealistic points of view and be rewarded for the results of our work with more means –knowing that is intelligence.

That is why I want to talk about matter, structured particles that interact and acquire emergent properties, properties that the individual particles forming the structure do not have; and about information, the judgments to matter that arise from a particular structural matter disequilibrium we call living beings or organisms. In living beings, matter and information exteract; this creates the possibility of closing the circle and directing the transformation of matter, taking matter beyond the random interaction and strategically into the service of us living beings. I also want to talk about the information that, alone, allows us unlimited travel in its domains and "to live" the perfect life –from a flat life, stoical, human, to a brilliant life, epicurean, beautiful, or if we prefer, an undulating one that moves between the human and perfect ideas.

Some tell me that I'm very theoretical and lead abstractions to the end. I hope to make the right abstractions, those that show that everything that happens to us is of this world, it is all natural, and nothing is supernatural. If something is missing, it is the understanding that the shortest path to a better world is more righteousness among us and within us.

My message is a message of justice. Everyone deserves to live and do it with dignity, giving to the community and receiving from it. It seems that there is no connection, among these ideas –matter, information, life– but there is. Information lets us, matter, attribute meaning to other matter, and choose our direction. Meaning is created from the depth of the living being structure.

info@matterinfolife.com

Introduction

"Every problem is a matter of information." Juan D. Arango.

The principles are simple. The principle of matter is the elemental particle. The types of particles are few and have two basic properties, mass and charge. There are practically a countless number of elemental particles that exist. The elemental particles are structured in different ways, forming groups we call atoms, molecules, cells, solids, liquids, planets, stars, etc. These groups are elemental particles that we call beings. Each being acquires properties according to the number of elemental particles that are involved and their structure and disequilibrium. As a principle of communication, let us define the words, being and object. A being is matter. An object is information.

The beings are grouped into two types, living beings and inert beings. In nature the living beings are differentiated from inert beings by a fact, their ability to replicate. When a living being replicates, it creates structures of elemental particles that have almost the same properties of the original structure. Traditionally, in addition to the principle (fact) of reproduction or replication, all living beings have been defined by other principles such as growth, adaptation and response to environmental changes, and death.

Those principles of living beings require a unique process, informational ability. Informational ability is the essence of life and allows the living being to create information when discriminating its environment. With informational ability and information, the living being makes decisions in order to conduct processes that differentiate it from lifeless beings. The living beings create groups through one process we call communication. Communication allows specialized activities between groups of living beings, which let achieve economies in the life processes. Communication allows a group of living beings to carry out projects that are not possible for any of them individually. The greatest project for a group of living beings is a community; a new living being.

The material principles behind the   processes of fire and life are the same. Surely they are a mystery to human beings. Both are natural processes of which we are starting to discover the principles. We know the material structures, we can define the components in each of these processes, but we do not understand the principles governing the actions that create light in the case of fire and information in the case of living beings. We human beings control our minds, but we need the teamwork of the cells, our living structure, for the completion of any project our mind decides on.

Keep an eye open when reading the book, your interpretation of what surrounds you is your information; if there are no living beings, there is no information. The book is divided into three parts, as the title: Matter, Information, and Life. In the first part we will look at structures of matter, as already mentioned, groups of elemental particles. The things we perceive help us understand the different groups of particles, because they allow us to create benchmarks and create conceptual structures. In order of elemental particles the first benchmark we have is the atom, then molecules, etc. The universe is at present [time], in disequilibrium, recreating the different structures of elemental particles. We will discuss a reality model where any group of atoms could be thought of as a structure, a set of interacting parts. Linking one structure to a system, the change in the structure creates a new system. The difference between the first system and the new system gives rise to the notion of a process, and the notion of interacting particles.

Having created the notion of a process from disequilibrium, the second part, information, will begin to define the process of creating information. This requires the imbalance of matter, not only to create any new structure, but also to create informational ability for the living beings. This whole process is conceptual; before gaining technical details, we need to understand the concept of information presented in this book. Then we can follow the same process that has lead to some discoveries, such as laser light: first the concept, then its implementation. In higher animals information can be accrued in magnetic fields from the brain, the mind. It is the information that is in use. But equally, there is information that is accrued with material means, which should be brought to the mind for processing, equivalent to the hard drive memory. It is memory that is in the brain cells and their connections. All the information that is in a living being is its particular truth. By sharing particular truth, conventional truth is created, which is information that we share. Without conventional truth, it is not possible to communicate among us. Then we will see ten basic information principles; there may be more, we can simplify those, they are a starting point.

In the third part, we combine the two concepts of matter and information; we explain the need of them for life. Paradigms are the basis of informational ability, the approaches or points of view on which we base our thinking. We analyze the concept of thinking as the movement of information. Thinking, informational ability, and so on, we have an orchestra-like process where our mind is the director. Information has two basic uses, to help us make decisions and to communicate among ourselves. Information exists at every level of the informational ability, in living structures. There are differences in informational ability among cells and multicellular beings, creating the notions of intelligence and consciousness. The notion of intelligence allows us to understand the parts of matter and the interactions between them, and the notion of consciousness helps us to understand wholes and our exteractions with them. We'll look at how an idea can be solid: reaching the decision point at the living being whose making it, building consensus among participants, and finishing as stated.

First Part:

Matter, Particles Interacting.

"It is generally believed that there is color, something which is sweet and something bitter; in fact there are only atoms and emptiness." Democrite

Matter is defined in a practical way as anything that has mass and charge and is occupying space. From a philosophical point of view, matter is defined as a being. Let me say that matter is any element that acts, that is, a being is anything tangible that our senses perceive directly or indirectly. In this book, or for the author, matter is a group of particles that has mass and charge, has a place in space and can create disruptive actions. Matter exists by itself.

When we speak of a group of particles, the idea of any figure comes to mind. Figures are structures of matter. Strictly speaking, we should not see a continuum, we should see elemental particles. We should see the small particles that form matter. But we are "trapped" by our size and the senses that keep us from discriminate matter at that level. Discriminatory senses perceive, discriminate, and communicate with our brain; then we interpret using our informational ability, creating a particular truth, our reality.

The universe has all the particles that exist, or the universe is all the matter that exists. Say there is a group of particles, specifically 10^{28}. What do we have in that group? A lot of proteins, millions of them or a brick wall, both options are possible. Today it is accepted that both of them are made of the same basic particles: protons, neutrons, electrons, and other less known particles. The same thing that happens to the protein or the brick happens to a plant, the sun, the moon, an animal, a quasar. All these bodies or beings are part of matter, groups of particles that have mass and charge, occupy a place in space and interact among themselves, forming operating structures we call processes. Each group of particles looks and acts differently because it has a different structure. With each regrouping, the groups of particles acquire emergent properties that are inherent to that type of structure. Not all particles are part of structures, but all are part of the universe. Those that do structures do so in countless ways and those that do not are ready to do so when the appropriate action or actions take place. Therefore, in the universe we have all the particles that exist, single or grouped, always acting and forming different structures, each with their own properties.

Let's look at some of these structures before entering a model to help us interpret our surroundings.

Chapter 1.

Structures

"Structures refer to mental models built after concrete realty. Furthermore these models are not obvious but demand an understanding of hidden, or deep aspects, of the matter at hand. Following this approach structuralism is an attempt to build models which can help understand or, as structuralists, would put it explicate the materials at hand". (Mark Glazer)

According to the definition of matter from this book, "a group of particles that has mass and charge, has a place in space, and can create disruptive actions." We may call those elemental particles' basic properties –mass and charge–, principles; and the results of all the particles and their interactions, universe.

Without the use of information, all that exists in the universe is particles. The only reference is the elemental particle, and everything is particles. By using information, we can say that a group of elemental particles that are not the universe and are combined according to their principles are a whole unit, or a system. The collection of particles can be identified by basic points of view such as the quantity, quality, and actions among them. Thus, we have an informational reference, and we may assign limits based on these three points of view. From quantity point of view, we may classify the groups of particles in the universe by using the number of elemental particles in the group. For this, we would use numbers and form groups of elemental particles having only one particle attribute, the number one. When we want to create other wholes from this mathematical point of view, say wholes forming a couple, we would say groups of particles of two units and assign them the reference number two, and so on. We would be using mathematics to classify the "properties" of matter. This concept should be meaningful to those who say that mathematics is the regulatory language or builder of the universe, as in this case it is the reference for the creation of groups. However, using a different point of view like quality, we could use the principle of the charge and say that if they come together they are a group that is attracted, and would classify them by attraction factor, saying they are either solid, liquid or gas systems. Using a different quality point of view, we may have another classification, and thus we can have as many classifications as we have points of view.

We humans, who are also immense groups of elemental particles, observe that based on the principles of matter, mass and charge, the elemental particles create basic groups with properties that allow us to distinguish one group from others. We call, the most basic group of particles, atom. Similarly, groups of elemental particles come together and form groups that we call molecules and more complex groups that we call planets or stars. In the same way, from other point of view, there are groups of particles that behave similar to our own and different from other groups. Like us, they can move around on their own, follow the light or move away from it; they are the groups of particles that we call living beings, and by their "independent" actions we can assure that the living beings handle (create/use) information.

Here we have used as reference the elemental particles. But normally we use one group of particles as a reference to other groups of particles. These groups of particles that serve as a new reference facilitate our discrimination of what surrounds us. When we use a group of particles as reference for comparison, we can call that group a part. In this way, the parts we refer to are between elemental particles and the whole. This is the case of an atom, which is part of a molecule.

In the intermediate-whole, or part, the original group of particles can gain, retain or lose properties. When the part is formed it acquires properties, the new properties of the intermediate-whole that we call emergent properties, there are not "found" in the particles that integrate it. The properties that remain through intermediate-wholes are called summative, they are added throughout the re-grouping; one of these properties is mass. Finally, there are some individual particle properties "lost" when these particles grouping into intermediate-wholes. Though, by changing the reference, we create groups of particles to form parts, and group of parts to form a whole, and so, by bringing together several of these wholes, we form a greater whole, the chain is a structure.

The reverse process is possible with the universe. By decomposing a whole, which we break down with some references that can be called parts or pieces, we arrive at elemental particles, which is the basic reference. Keep in mind that the references are decided, they are information; humans create references very well. The nature of matter is particles; one part is an arbitrary reference, established to facilitate the understanding of what we see. Parts are common particle groups, which are conventional references of the subject matter; a planet is a part of the solar system. The groups of participles which do not follow a standard pattern when disassembled, like a half of the Earth, we can call pieces.

In summary, there is one whole, the universe, which is formed by all elemental particles. They form two extreme references. Humans have created scientific references to discriminate and better judge their surroundings. These scientific references help us create structures that we use to describe systems and processes. Scientific references are usually temporary because new scientific references bring new technology represented in new tools, which in turn brings new discoveries serving to continue the improvement cycle. Then what lies in between elemental particles and the universe is the way we create references to build structures, but in each case, these references are an issue of information. Structures do not depend on what we think; they depend upon the properties of each particle in the structures and the collective emergent properties of them in the structures where they all combine. The combination of all existing particles is the universe.

Let's look at some structures with different combinations of elemental particles that serve as a reference to discriminate the universe.

The Structure Of Parts Of Matter.

Let's say this is the structure of matter studied by chemists. In this structure, the baseline is the atom. The atom is a group of elemental particles which is arranged on the basis of those particles called protons. The classical atom has a nucleus and a structure of particles moving around the nucleus. At the core there are two types of particles, neutrons and protons. Moving around the nucleus, there are particles called electrons, which have many types of orbits. The different structures of these elemental particles are responsible for the creation of chemical elements. Depending on their structure, they can have properties such as those of gold, silver, diamond, etc. This tells us that, according to their grouping, these particles may have different emergent properties. This structure, studied by chemists, has more than one-hundred basic combinations or "normal" combinations, we said, taking the proton as a reference. Another interpretation is that the structure is defined as normal if it has the same number of electrons as it does protons, which makes the structure neutral considering that the protons have positive charges and electrons have negative charges. Today, there are one-hundred-eight different structures of atoms defined with clearly distinct parameters by chemical scientists.

When speaking of normal atoms, we are defining the atomic structures, and this leads us to believe that there must be abnormal atoms, those that do not follow the proton and electron rule. Indeed, there are over one-thousand variations of the abnormal atoms, because they are missing or have too many neutrons or electrons versus the number of protons, which we said, is the basis for classifying this structure. That is not the same with changes in the number of protons, since the number of protons is the reference. If you change the number of protons in an atom the atom has another name; it is another chemical element. Let's look at an example. The normal hydrogen atom should have a proton and an electron. If it gains a neutron it becomes an isotope of hydrogen, 2H (one proton, one neutron and one electron), deuterium or D, an "abnormal" hydrogen atom. If it gains a proton it becomes an isotope of helium, 3He (two protons, one neutron and one electron) or helium-3 and it is an "abnormal" helium atom.

Normal or abnormal atoms have other characteristics that are given by the levels electrons occupy, which are called energy levels. The number of electrons in the outer energy level creates a property, the electronic valence. The electronic valence defines the inclination to interact with other atoms more or less easily. In other words, the electronic valence theory determines how easy it is for some atoms to combine with other atoms or themselves to form molecules.

The molecules can have from two atoms, as the molecule of table salt, NaCl, to hundreds of them, as proteins and other organic molecules. There are very large molecules, such as those of the Polyvinyl Chloride (PVC) from which plastic pipes are made. From a structural interpretation, the molecules are groups of groups of particles. Atoms are groups of basic particles and molecules are groups of atoms. The molecules form another level in the structure of particles of which there are millions of combinations. Under the new structure, molecule, there is a change of properties, that is, the combination of atoms also acquires emergent properties. The molecules have different properties than the atoms that make them up. An example: water, H_2O. The hydrogen atom is very unstable, it easily combines with itself. The oxygen atom is very unstable; it combines with most other chemical elements (Periodic Table of Elements). In combination, an oxygen atom with two hydrogen atoms is water, which has little tendency to recombine or is quite stable. This tells us that the ability of oxygen or hydrogen to combine is substantially reduced with the new structure.

Normal and abnormal atoms, molecules and other particles can still, according to their energy levels, restructure to form combined mixtures thereof. These mixtures acquire a characteristic that we call the state of matter. There are three normal states: solid, liquid and gaseous. It is noteworthy that at these levels, we do not require the help of instruments to identify the structures, that is, these groups of elemental particles are easily perceived by our senses or by their effects on us living beings. Gases are more difficult to detect by our senses and are responsible for the atmospheres on planets like Earth.

The basis of the modern structure of matter has changed, not in its essence but in its form. In today's view, the atomic structure reference quarks as particles structure to form protons and neutrons. It is said that quarks are the size of electrons. Quarks add an additional level to the classic structure of an atom. A modern interpretation of the structure of an atom is that there are new groups of particles forming the nucleus; another interpretation is that there are particles smaller than protons and neutrons in the nucleus. This new idea does not change the concept of atoms from the point of view of chemical science, which basically maintains the same structure designed in the periodic table of elements. Also, among the atoms, the interaction principles to form chemical compounds are retained. In other words, regardless of its true structure, the concept of atoms and particles will survive in chemistry. The question is: when we learn more about the groups of particles that form the atom, will we be able to further explain the structures of life?

This section shows how the world is made of very small particles that are grouped and regrouped, creating structures according to specific levels. Let us say that a structure is formed by grouping parts, and each new group of parts acquires properties specific to this new group of particles, which we call emerging properties, as they result from the new group of particles that was formed. "At each level, structure dictates the properties of the group" (Arango C. , 2011). The new being, a whole of parts, has new properties that emerge with the new group, and some of the properties of the parts are lost to the new structure while others remain. In summary, the fundamental particles create atoms and atoms create molecules. Then you can use elemental particles or atoms as a reference for molecules. Combinations of atoms and molecules create and are a reference for crystals. There will also be other more heterogeneous groups of particles –solids, liquids and gasses– that are defined by their energy level, referred to as enthalpy. We can see this structure of matter as particle aggregation, synthesis.

Stellar Structure Of Matter.

Let's say this is the material structure studied by astronomers. This structure of matter has had more historical meaning to us than the previous structure. Some of the participants in building it have paid with their lives to maintain respect for their observations. Let us describe the structure of matter from the heavens, stellar bodies, material structures larger than us that allow us, humanity, to have our platform, the basis for transactions that does not "move." The platform that is our immediate environment, serves as support, and lets us share with other living beings, is Earth.

Let's look at this structure of matter as a whole disaggregation, analysis. The universe contains all the matter that exists. The universe is made up of groups of galaxies; these galaxies are composed of stars and equivalent celestial bodies. The stars are comprised by planets and planets are comprised by their moons.

These structures also have emergent properties. We are considering a basic model of stellar bodies where the magnetar, the pulsar, the black hole, and the quasars are taken as stars, and comets are taken as planets. With the basic model, we can consider the normal stellar structure as one that has cycles governed by gravitational forces. Simple cycles are those of the moon around the Earth and the Earth around the sun, including the sun in the galaxy, etc. There are stellar bodies that do not have clearly defined cycles. And from there, stellar structure can also be declared as "abnormal" where cycles are not defined or are not easily described. Examples of this are meteors that have hit the Earth or other planets, and the case of the comet Shoemaker-Levy 9 that collided with Jupiter. Also, some astronomers predict the collision of our galaxy, the Milky Way galaxy, with Andromeda.

In order to understand or see things from the perspective of human beings' informational ability, let us review how this "simple" view of the stellar structure of matter was developed; this point of view of the universe has conflicted humankind many times over. Let me take you to what Wikipedia says about the history of astronomy (Wikipedia).

> *"From immemorial times, man has been interested in stars, they have shown constant and unchanging cycles during the short period of human life, which was a useful tool to determine the periods of abundance for hunting and gathering or those like the winter than was required of preparing to survive an adverse climate changes.*
> *The practice of these observations is as certain and universal that has been found throughout of the world in all those parts where man has lived. It follows then that astronomy is probably one of the oldest professions, manifested in all human cultures.*
> *The immutability of the heavens, is impaired by real changes than the man in his early observations and*

knowledge could not explain, from there the idea that in the sky dwelt powerful beings that influenced the destiny of communities and human behavior had worship therefore required to receive their favor or at least avoid or mitigate their punishments. This religious component was closely related to the study of the stars for centuries until scientific and technological advances were clarifying many of the phenomena at first not understood. This separation did not happen peacefully and many of the ancient astronomers were persecuted and judged to suggest a new organization of the universe. Currently these religious factors went to modern life and survive as superstitions.

Today we know than we inhabit a tiny planet in a solar system ruled by the Sun that moves in the first third of its life and is located on the outskirts of the Milky Way, a barred spiral galaxy consisting by billions of suns, it possesses as other galaxies a super massive black hole at its center and forms part of a local galaxy group, which, in turn, is in a super-cluster of galaxies. The universe is made up of billions of galaxies like the Milky Way and has been estimated an age between 13 500 and 13 900 million years, and its expansion is accelerating constantly."

There were many humans building this model, Aristarchus of Samos' (310-230 BC) created the heliocentric model, with a static sun. Ptolemy's (100-170) geocentric theory, received strong support from Aristotle and his school, reign more than a thousand years. Nicholas of Cusa (1401-1464), who in 1464, suggested that the Earth was at rest and that the universe could be conceived as finite. Nicolaus Copernicus (1473-1543) re-incorporated the helium-centrist ideas. The pair of Tycho Brahe and Johannes Kepler (1571-1630) the former made observations, and the later using mathematics, finally reached the understanding of planetary orbits with ellipses testing rather than the perfect models (circles) of Plato and could then state his laws of planetary motion. Galileo Galilei (1564-1642) was one of the strongest advocates of the helium-centrist theory. He built a telescope from an invention of Dutchman Hans Lippershey. It is worth mentioning a martyr of astronomy, Giovanni Bruno, who spoke of the infinite universe and described the path to wisdom (Summary from Wikipedia). To complete these comments, let us note a test of the limits of the universe and the informational limits that are imposed upon us; the picture, Hubble ultra deep field, taken by Hubble telescope. When positioned to a single point for 48 hours, that is, taking a picture that lasts 48 hours, countless galaxies and stellar bodies emerged, which were never before seen and are invisible to the naked eye, because it is not allowed by our material structure.

This section shows the structure of matter in stars, another interpretation of the same matter of the universe. We started in the greatest whole, and subdivided it. Bearing in mind the above structure, we find that we are in a place between the very large and the very small. It must be noted that the stellar structure is as much a part of our environment as is the structure of the parts of matter. Each one has its own influence on us, but the magnitudes are very different because of the forces and the distances. Seeing each star as a unit is a great abstraction. Assigning personalities to them and asking them favors both requires an outstanding informational ability only seen in human beings.

Living Matter Structure

Let's say this is the material structure as seen by biologists. This structure of living beings, you, me and others, is a structure of matter like others previously mentioned. We are talking from the point of view of the structure. As mentioned, elemental particles form atoms, atoms form molecules, and the molecules of life are no exception. The most famous molecule structure, the genome, has millions of atoms composed of elemental particles themselves. For the time being, let's look at the structure of matter, which is responsible for life.

The structure of living matter begins with a "selection" of atoms, not all the one hundred eight types of atoms are involved in this structure. Interestingly, to say the least, three types of atoms: Oxygen, Carbon and Hydrogen (O, C, and H) represent over 96%[2] of the matter that exists in the human body structure. Adding three atoms –Nitrogen, Calcium and Phosphorus– to set six types of atoms (O, C, H, N, Ca, P), achieves approximately 99% of matter in the structure of the human being.[3] The remaining 1% includes over twenty different elements. Within the classical structure of life, we emphasize that the atoms and groups of fundamental particles that make up living beings are considered dead. Referring to the basic concept that gave rise to this book, without leaving the classical concept of life, the combination of different atoms and molecules in a cell (or its parts) creates an emergent property, which is informational ability, the essence of the existence of living beings.

Within the classical structure of living matter, the cell is the smallest living creature, the *atom* of life. The atom of life is commonly called: prokaryotic cell. It is organic matter containing cytoplasm and other structures called ribosomes, including DNA, in addition to other organic molecules, all enclosed by a plasma membrane. In

---

[2] It is based on mass percentage. For some not trained in engineering, mass is not a clear concept. Mass is responsible for weight, but is not the weight. Weight is the force that a body, a group of particles, exerts on another. Conversely, the other body, the latter exerts on the first. As an example, if you weigh 98 kg on Earth, which has a gravity of 9.8 m/s[2]. You weigh 16.2 kg on the moon which has a gravity of 1.62 m/s[2]. Keep this in mind; we have become familiar with many facts. A box of cotton weighs far less than one of iron. But a pound of cotton has the same mass of one kilogram of iron and therefore the same weight here on earth.
[3] Independent from that high percentage of these elements in living beings, it has been discovered that there are other chemicals that were required for the functioning of multicellular beings.

these cells, there is no clearly defined nucleus membrane (Dupage). Formerly, all elements within the prokaryotic cell were defined as protoplasm. Types of these cells are bacteria such us E. coli.

Doing aggregation, we have the eukaryotic cell. The eukaryotic cell has several equivalent structures –prokaryotic cells– from which the nucleus and mitochondria are the most renowned. According to this structure, the eukaryotic cell is equivalent to a *molecule* of life, made up of a group of prokaryotic cells, atoms of life. That is, the eukaryotic cell has membrane-bound parts, each of which can be assimilated to a prokaryotic cell[4] and are technically called organelles[5]. There are still discussions about whether eukaryotic cells are a group of prokaryotic cells, but some authors see it that way. Within this structural concept, it is widely accepted that mitochondria, one organelle of eukaryotic cell, is a prokaryotic cell that shares features with the eukaryotic cell in a symbiotic manner, providing ATP and receiving food. This symbiosis has matter interchange as well as information interchange. The nucleus is the most recognized organelle in the eukaryotic cell. All these structures create emergent properties, where the eukaryotic cell is more specialized than the prokaryotic. On average, eukaryotic cells are ten times larger than prokaryotic cells.

Eukaryotic cells form multi-eukaryotic structures. These structures are called superior living beings and form three branches: animals, fungi and plants. The branch of animals includes humans. To talk about one branch, animals, their structures have groups of cells that represent organs and are associated with colonies. The brain as a group of cells or colony has an emergent property, the informational

---

[4] This statement might be controversial, but can be justified by the characteristics of the organelles that reflect many of the characteristics of prokaryotic cells.
[5] These groups of parts of eukaryotic cells are equivalent to the organs in multicellular beings; where specific functions are accomplished for the cell.

ability of animals that creates a whole from the systemic point of view, a self. The creation of this self has many connotations, one is accepted that the brain has a psychic unit, which is known as the mind, emergent and created by the synapses[6] of neurons.

To complement the structures of human beings[7], we can say that the structure between eukaryotic cells and human beings can be seen as having two additional levels, organs and systems, which have informational structure. The cells form organs like the heart, but the heart does not make sense without veins, arteries, and capillaries. All together, they form another level in the structure, which is the circulatory system that carries blood to all parts of the body and connects with other systems to support humans' vital functions. These two levels, organs and systems, can be considered support for the operation of the structures of human beings, they are informational.

Continue structuring, we have room to say there is still a larger unit, a group above the level of human beings: humanity. Humanity is a community of living beings on Earth, which is also made with similar principles to the ones mentioned of other living beings' structures. Here, there are means such as culture for information integration, and commerce for matter exchange, turning human beings into humanity. Then humanity is a higher level living being than a human being.

Of course, it is possible to define other structures under other points of view; you might have a structure for nature that has a rich

---

[6] In the brain there are hundreds of thousands of synapses per second, those electrical signals are capable of forming magnetic fields which are read by electrodes that are installed on the scalp or forehead and can be registered in different ways.

[7] You can also include many other multicellular creatures. Remember that these structures are a reference the book does not expect to cover them all, nor all the details on the structures. This is a book on information not a science encyclopedia.

history, is more scientific, and has long been familiar, so there is a reason not to accept these structures. Let us just pretend to outline structures of matter from the viewpoint of a chemist, from the viewpoint of an astronomer, and from the viewpoint of a biologist. This last structure, the structure of life, is a special structure because it makes use of information.

It is worth mentioning again that our senses limit us. No one has seen an atom nor as we said, the universe displayed by the Hubble telescope. We see the results shown by the tools or equipment that we built in a unique way. It is unique because it has not been made by any other living creature on Earth, to the extent that we have come to believe we are beyond the nature of matter. This happens when we classify what happens in nature; we have heard of natural structures, made by nature and the artificial, man-made ones. What vanity! Are we not part of nature? Are other living beings not changing nature?

Reaching these three structures of particles has required a lot of material and informational work from many human beings: numerous projects to create tools, methodologies, and certainly inventions –mental model creations– that supplemented or entirely rejected formerly renowned tools or methodologies. In the next chapter we will discuss disequilibrium, which allows us to explain many ideas, particularly help us understand that without disequilibrium we could not exist.

Chapter 2.

Disequilibrium

"The most obvious and fundamental phenomena we observe around us is movement." Alonso & Finn

In the previous chapter, we talked about the structure of matter in a static way. But matter is active throughout its entire structure, among levels and at every level. These actions create changes in one or more levels of the structure; these changes can be seen as movement, which can be considered as the operation of the structure. Actions on structures of matter are defined by the characteristics of elemental particles, their mass and charge, and then by the emergent properties of these structures, where there will be different types of actions for different groups of particles. In this aspect of material actions of matter, let have a quote from university physics book that makes an interesting overview of "different" types of actions among matter, which its author calls matter interactions.

> *"Some questions naturally come to mind when we think about the particles that make up our world. Why and how are electrons, protons and neutrons bound together? Why and how are atoms bound together to form molecules? Why and how are molecules bound together to form bodies? How does it happen that matter aggregates itself in size from small dust particles to huge planets, from bacteria to this marvelous creature called* Home sapiens? *Why do all bodies fall toward the Earth, while planets revolve around the Sun? We may answer these fundamental questions, in*

*principle, by introducing the notion of interactions. We say that the particles in an atom interact among themselves such that they produce a stable configuration. Atoms in turn interact to produce molecules, and molecules interact to form bodies. Matter in bulk also exhibits certain obvious interactions, such as gravitation, manifested for example in what we called **weight**.*

One of the primary objectives of the physicist is to disclose the various interactions of matter and to express them in a quantitative way, for which mathematics is required. Finally, an attempt is made to formulate general rules about the behavior of the matter in bulk which result from these fundamental interactions.

*So far, four kinds of fundamental interaction have been recognized: **gravitational, electromagnetic, weak** and **nuclear**. Each is related to a particular set of properties of matter and phenomena. The gravitational interaction is the weakest of the four. However, it is the most easily recognized force since it is manifested by an attraction between all matter. It is responsible for the existence of stars, planetary systems, galaxies and in general all large structures in the universe; it is related to that property of matter called **mass**. The electromagnetic interaction is related to a property called **electric charge**. As we indicated before, it is responsible for holding atoms, molecules and matter in bulk together. The weak interaction manifests itself through certain processes, such as some kinds of **radioactive disintegration** or **decay**. It is related to a property called 'weak' charge. The strong or nuclear interaction holds protons and neutrons together in atomic* *nuclei as well as quarks inside protons, neutrons and pions...*" (Alonso & Finn, 1992)[8]

Disequilibrium exists; it is a part of the nature of matter, actions among particles. However, this disequilibrium is limited by the properties of the group of particles involved. If the disequilibrium were total, if it had no limits, particles would not bind to one another; all elemental particles would be free; there would be no structures. The fundamental particles would be loose; they would be like gases. Then there must be some affinity among the fundamental particles to attract one another, and in turn another characteristic, which we can call unaffinity, which by counter-logic causes particles not to compact. Otherwise, the structures would stiffen and we would only have solids, perhaps one big solid. In other words, something attracts elemental particles so they structure, and something polarizes them so they do not collapse or lose their movement. The essence is that an action is needed in order to maintain structure, and another action is needed to maintain disequilibrium; something like action and reaction or attraction and repulsion. By now, think that mass and charge principles could be enough to maintain matter disequilibrium –mass with its inertia, and charge with its action at distance.

Let us mention the following sections on movement to maintain a sequence that ultimately allows us to understand the concept of matter and information and the combination of them in the process

---

[8] To show the difference in magnitude between these interactions see the figures expressed with all zeros and with reference to the most familiar to us, the gravitational interaction has magnitude = 1.

With gravity equal to 1, the earth has magnitude interaction about 10.

The weak interaction has magnitude interaction of 100 000 000 000 000 000 000 000 000 000,

The electromagnetic interaction has magnitude interaction of 1 000 000 000 000 000 000 000 000 000 000 and

The strong interaction has magnitude interaction of 100 000 000 000 000 000 000 000 000 000 000.

of the living beings. Understanding the dynamics of the structures is as important as understanding the structures themselves because the dynamical properties emerged from the structure. We may have two "identical" structures, but their dynamical properties will not be the same because these structures are in different places and then subject to different actions from their environment.

Disequilibrium Of The Parts Of Matter

The structures of the parts of matter, as we have said, begin with fundamental particles. These elemental particles are in constant motion. We cannot detect single elemental particles that make up the structure of matter, so in the same way, we cannot detect the movement of these fundamental particles when they move from one position to another, transforming the structures, creating change. There are several reasons for that. One is the quality of change, another is the difference in magnitude or levels, and another is the large number of simultaneous actions. Therefore, there is constant movement in the structures of the parts of matter, but it is not easily detected.

As it was already mentioned, we are trapped by our senses. We need tools to go beyond the limits of the senses and discover at least the end of the parts of matter, atome. Let's use the word atome to separate the basic system of chemical elements, atom, from the idea of indivisible particle, the basic unit of matter. The dynamics of fundamental particles is being studied by quantum mechanics; its scope is still very limited, and we are in the process of better understanding them. Hopefully, soon we will have advances in the knowledge of the parts of matter, their structure and their dynamics. In particular, the dynamics of parts of matter –elemental particles– which are essential to understand more about the properties of living matter, require more significant research.

Disequilibrium Of Stellar Matter

The disequilibrium of stellar matter is more familiar to us, it is usually cycles. It is explained by known laws of physics. But there are aspects of many stellar bodies that are a mystery to us, issues not related to the dynamical properties among stellar bodies, but more related to the structures of the parts that make up these stars. In other words, stars can give rise to other types of movement among the parts of matter that form them. That is, given its large size, the gravitational attraction in turn affects the lower levels of the structure, the properties of its internal structure, changing the very basic dynamic of structures of the atoms that compose these stellar bodies.

One example is the black hole. This star has such a mass that it catches the light passing at some distance from it. At some point, before being a black hole, this star had some atoms with "normal" properties; it was a normal star. The structural change from normal star to black hole not only affected the normal star, but it also affected the groups of particles that formed the star and now make up the black hole. According to astronomers, the density of the black hole has closed the distance between elemental particles and possibly stiffened its atoms. This would be an effect of the particles structure on itself.

Disequilibrium Of Living Matter

The disequilibrium of living matter, not related to information, is given by different properties of fundamental particles. One of the most fascinating aspects of this structure is repetition. The structures of living matter are repeated almost identically representing systems, and play in dynamics that have nearly identical properties, representing processes. There is important to remark, as we will see in levels of informational ability that there is a clear informational structure, which does not follow the same material disequilibrium, but does follow an informational one. There are many documented cases of dynamics of the living matter on many fronts. None of these explains the dynamics through which organisms –living beings– discriminate or judge. That is, they do not explain the way in which the informational ability of the living beings is established. Today, the most common explanation is that informational ability comes from the stars, or at least, is supernatural.

Chapter 3.

One Model For Nature

"All models are wrong [models of reality], but some are useful."
George Box.

The previous chapters have shown two material facts, structures and disequilibrium, which create reality. Here, we call it a model for reality. The model for reality or simple reality is the attribution of meaning to matter. This attribution of meaning itself is information. Now, there is also a biology model that names reality: nature; and splits it in two facts: live matter and inert matter. The facts for the living are birth, growth, reproduction and death. But under the reality model, the facts for life are structure and disequilibrium. Then what is the difference between live matter and inert matter? The main difference between live matter and inert matter is the ability to direct disequilibrium. This ability is born from the informational process of live matter, which we call informational ability. Informational ability comes from a material structural disequilibrium, forming a dynamical equilibrium toward preserving the material structural disequilibrium itself. Here we call the fact of directing the material structural disequilibrium informational ability. The informational ability is responsible for producing information. This structural disequilibrium is a synthesis of simultaneous actions among matter at different structural levels.

The property of life emerges from a synthesis among material structures and their disequilibrium. The concept of life, like the reality concept, is debated. The greatest disagreement about life comes from its origin. Does life come from outside of the structure or does it come from within the structure. The point is worthless in front of the synthetical fact of life. We, living beings, are an example of material synthesis; we are synthesis of various structures in various levels of disequilibrium. Now let us conceptualize how informational ability translates reality to nature.

From the point of view of reality, life directs matter. Informational ability is the way to choose a direction for matter. To choose direction is a process; a decision is the result of the process; a decision is information, which in turn, is a creation of meaning. Living beings, as matter with informational ability, work to transform material structures to a chosen structure.

All that exists are elemental particles in act. Elemental particles are the principle of every element in reality; one particle or groups of them are beings. Beings, because of their interactions, have summative properties and emergent properties. The summative properties of elemental particles come from the quantity of particles. More elemental particles bring more of the same property. The emergent properties of elemental particles densities come from their new structure and their disequilibrium. If you break the structure or the type of disequilibrium, the emergent property is lost. Emergent properties are not found in the elemental particles; they have elemental properties themselves. Some emergent properties can be found on sub-groups of elemental particles (beings) that form a bigger structure. The emergent properties are responsible for the occurrence of life.

According to scientific models, elemental particles' actions have two characteristics which are responsible for incessant motion: inertia and action at distance. These elemental particles have two basic principles/properties: mass and charge, which are responsible for particles' actions' characteristics. All elemental particles integrate into one group: matter. Matter has an uneven distribution of elemental particles that creates different matter densities or particles' densities. The particles' densities attain emergent properties according to their densities and the actions among them. It is important to understand, in contrast to the scientific model, that matter densities synthesize all elements responsible for the structures and the disequilibrium these structures have, as well as the source of living beings.

So some particles densities are alive or dead. Our model for nature sees one emergent principle in living or alive densities: informational ability. The informational ability lets living densities attribute/create meaning to/from other densities' qualities and decide to act or not to act over other particles' densities. The attribution/creation of meaning through informational ability is what we call information. We can say that living densities create information when they create meaning. Let us call the attribution/creation of meaning to/from a being an object.

To bring awareness to our size relative to those elemental particles, and to help us understand our complexity level, let us compare the size between the diameter of an Iron atom and the Earth. Let's also have an intermediate reference, one small rod 6.3 cm in length. When comparing the atom to the rod, we need two hundred three million (203,000,000) Iron atoms (Fe, covalent diameter 310 picometers) in line to have the rod 6.3 cm in length; and in turn, when comparing the rod to the Earth, we need two hundred three million (203,000,000) rods, each 6.3 cm in length, to get the Earth's diameter, twelve thousand seven hundred and fifty six (12,756) kilometers.

This relative size shows the amount of elements that build us, an estimated 10^{28} atoms or 10^{14} cells, which contribute to the difficulty in understanding nature and the creation of the living. The complexity belongs to the amount of elements involved, not to the principles involved in the elemental particles or their actions. In other words, matter principles are simple –mass and charge, and inertia and action at distance. The complexity emerges from the amount of particles and their simultaneous interactions.

Material limits do not exist until we reach the elemental particles themselves. To grasp the concept of densities/limits, let us use atmospheric pressure maps as an example. The atmospheric pressure maps show high pressure areas that coexist with low pressure areas. The map has arbitrary lines that indicate different pressure areas; these lines are called gradients. Gradient lines do not exist in nature but are used to present weather conditions. Keeping this example in mind while reviewing our model of nature, we observe a similar phenomenon; high density areas coexist with low density areas, but this time, very high density areas abruptly change to very low density areas, there we perceive (draw) lines, which are informational limits. An example of this density changes is the case of Earth and its atmosphere, where a high density area, Earth itself, transitions to a low density area, the atmosphere, and we see clear limits between the surface of Earth and the atmosphere.

Then, at our size or level, the abrupt changes in densities look like very clear limits. But observing this from a smaller size or lower structural level, the clear limits start becoming much diffused, and there is no observable clear limit because there are spaces among the particles and they move through what appear to be tangible "limits." We can abstract that at that level it would be as if we were looking at the heavens; we would "see" those particles as if they were stellar bodies, but there would be no light; we would only have elemental particles moving in the dark.

Philosophical Point Of View
Let us explore matter and information, and structure and disequilibrium, from the philosophical point of view. Using philosophical terms of 'existence and essence', we can see particles as existence and the attribution/creation of meaning -information- as

the essence. Matter exists; living beings attribute meaning to matter densities, creating essence, which we later discuss as objects. The essence, the object created, could be different for every living being since each living being has its own point of view. An example is the object that a blind individual can create when introduced to a picture, versus the object that a "normal" person creates.

There would be no essence without attributing meaning to matter or creating meaning from existence. We create fear, rage, love, duty, etc. with our information ability. In both cases, attribution/creation of meaning represents the notion of information, one from the concept of observer and other from the concept of actor. The observer attributes meaning to the behavior of the actor, the actor creates meaning and plays –behaves on it. Sometimes, the actor can observe himself acting, which is a deeper informational ability.

Classical metaphysics asks: "What is there?" The answer is elemental particles in act. The elemental particles' densities actions include informational ability, responsible for the creation of information. Information is the attribution of meaning to particles' densities structures. Informational ability is obvious; nonetheless, it took a long time before Descartes highlighted it in his famous saying: "I think, therefore I am." In our model of reality, this goes backwards; I am, therefore I think. First the existence, matter, then the essence, structure; since in order to perform any action you need to exist. You exist as a structure in disequilibrium then you reaffirm the attribution of meaning to a material structure –information.

In the substance versus accident dichotomy, we can see substance as the material structure in front of us and accident as the disequilibrium responsible for the different matter densities' changes. Matter densities at the elemental particles' level are flowing, transforming themselves according to matter principles. Matter is grouping into different densities and we observe those changes and give meaning to the new densities as accidents. To better understand what I'm saying, let us perform an exercise. Look at the palm of one of your hands for a moment and then close your hand into a fist. If your hand is the set of particles in a specific system, then your hand, open or closed is still a hand. From the philosophical perspective of its substance, the elemental particles are all there. But an open hand is different to a closed hand; they are two different accidents of the same substance, one system when your hand is open, and another system when your hand is closed. The accident changed -form; the substance remained –elemental particles. Again, we have material structure –in this case the matter– as substance, and the disequilibrium as accident.

The 'universals versus particulars' dichotomy can be pair to matter and information concepts. Matter is particular. No two things are equal. But through abstraction's point of view, distinctions are made, abstracting properties and obtaining universals. We obtain equality under the parameter abstraction, i.e. simplifying reality the way that we define it, creating objects. When we invalidate all parameters of matter, we perform the highest abstraction and create the universe concept, from which we derive the idea of the unit. Material structures are particular. Examples are flowers, pears, atoms. Each one is particular; they have unique count of atoms, light reflection, shape, etc. We living beings, through cultural and scientific conventionalism, assign properties, creating equality according to the properties assigned to beings. Those groups become universals (objects), many beings sharing same properties (abstractions). Those groups of flowers, pears, atoms, are universals because they share the properties we assigned. Then, objects are universals for the groups of living beings that accept them as representations of those groups of beings. Beings are particulars since there are no two equal beings.

'Abstract and concrete' create another dichotomy. Again, in order to create information, we abstract characteristics from the matter densities -beings- and produce ideas. Matter -beings- are concrete and information -ideas- are abstract. As we will discuss later in the book, when creating meaning straight from matter densities we create shapes. When we re-create those shapes we compare them and create more abstract concepts, pure information concepts. That is the case when comparing two separate circles; we decide there is room between them. The space between them is an abstraction, since the space is only a concept coming from extrapolating the existence of the circles versus the abstraction of the space, which do not have any concrete element, there is no matter there.

Now, let us discuss the dichotomy of 'determinism and indeterminism'. The future, as the past, does not exist. The future is an informational ability creation, a model. The past is memory, accrual of information that we create at present when we are seeing, hearing, etc. With our accrued information and our informational ability, we extrapolate the future; we imagine what is going to happen. But what is going to happen is the result of all elemental particles interactions, how can you, I, or anybody else possibly track them all? This is regarding the future of the universe. But, for us living beings, we imagine our future by creating plans. The issue is that other living beings or matter densities can ruin our plans –this excludes each of us, when for a reason or emotion, we act and change our plans. Obviously, what is going to happen, will happen, and you are free to call it determinism. Now, since we only control our attitude (mind) we do not have control over many of our internal actions and less over the externals ones (disequilibrium), these create indeterminism. Bear in mind that most living beings are bound to a main objective: to preserve their lives. This in turn is informational determinism, karma, etc.

Finally, in this dichotomy discussion, we could see matter and information creating one dichotomy. The dichotomies themselves are information, objects, attribution of meaning. Reviewing our dichotomy discussion, we lose the integral point of view, the combination of both dichotomic terms. In most cases, we discuss them as if they cannot exist together, but the two dichotomic terms are bound to the elements mentioned in this book –matter and information. The missing issue is their integration. When we integrate both dichotomic terms –matter and information– we discover one whole: the living beings.

Physical Point Of View

The material structure and its disequilibrium are represented in the scientific literature with two points of view: "what is this thing?" and "what does it do?" (Ashby, 1957). The former question takes us to The General Theory of Systems (structure). The later question takes us to The Cybernetics Theory (disequilibrium). These models are used in almost all areas of science.

Those models can help us understand our model for nature. Let me oversimplify how these models work. When we use the general theory of systems, we focus on structure. In the first place, we define *limits*. Secondly, we make *comparisons* to gather common elements. Later, we *generalize* results over the different structures we analyze; we define what the components of the system are and how they relate.   Using the cybernetics model, we focus on disequilibrium. In the first place, we define *states* for a system. Secondly, we make *comparisons* to gather common elements from those states. Later, we *generalize* results over the different states and find a behavior for the system.

To explain the systems concept, Bertalanffy used a simple table; see Figure 1A –System Concepts Illustration. The table has six cells presenting three comparisons, each between two groups. Each comparison helps define one of three basic concepts on how to see reality through the systemic approach (Bertalanffy, 1976).

Let's look at the comments made about Figure 1A. The first step is to define limits. Once limits are defined, he starts comparisons and draw distinctions.

	System A	System B
Comparison 1	o o o o	o o o o o
Comparison 2	o o oo	o o o ●
Comparison 3	o-o-o-o	o-o |x| o-o

Figure 1A. System Concepts Illustration.

Comparison 1 shows two systems –A and B– which differ by the *quantity* of their elements. The distinction term used by Bertalanffy is *number* rather than quantity.

Comparison 2 shows two systems –A and B– which differ by the *quality* of their elements. The distinction term used by Bertalanffy is *species* rather than quality.

Comparison 3 shows two systems –A and B– which differ by the *interactions* of their elements. The distinction term used by Bertalanffy is *relations* rather than interactions.

Keep in mind that we are not talking about each of the circles individually. We draw distinctions between wholes, not the parts within the wholes. That is, we talk about all the elements at once; the focus moves from each circle into the group of them, which is the notion of a system. In this analysis we are creating meaning or remembering previously created meaning learned at school, home, etc. The conventional meaning or properties –number, species, and relations– help us communicate and integrate communities.

Let us see the structure that exists within Figure 1A. We start *deciding* over subject matter limits we are going to use. We can start seeing each circle as an elemental particle, *one sub-system*. We can see the group of circles in each cell's table –system x, comparison y– as a whole, a matter density, *one system*. All six groups of circles in Figure 1A can be seen as a table, matter, the universe, *one meta-system*. With all wholes –twenty five sub-systems, six systems, one meta-system– we create a chain of wholes, *one structure*.

When we look at both ends of the comparisons –1, 2, and 3– at the same occurrence, systems A and systems B are both in front of us at the present time. Equally important, we are creating meaning from the distinctions when comparing them: quantity, quality and interactions. Finally, we could say that the systemic model brings attention to the structure of matter.

The cybernetics model notices the disequilibrium, focuses on movements or patterns of movement. When grouping all the movements of the system we get its behavior. For our model of reality, the behavior is called process. Although Bertalanffy uses the illustration, Figure 1A, to compare features among systems –such as those mentioned of numbers, species and relations– let us work the cybernetics model from there, to "explain" the elemental concept of disequilibrium. To make the task easier, let me recreate Figure 1A as Figure 1B. Cybernetics Concepts Illustration. First, we need to rewrite the title of columns from system to *state*. Then, Figure 1B, will have six groups of circles –state x, comparison y– as wholes and three comparisons with two states each.

	State A	State B
Comparison 1	o o o o	o o o o o
Comparison 2	o o oo	o o o ●
Comparison 3	o-o-o-o	o-o ⌐x⌐ o-o

Figure 1B. Cybernetics Concepts Illustration.

Now let use our imagination travel between states. Grab 'state A, comparison 1' (o o o o) as the present for comparison 1; at this moment 'state B, comparison 1' does not exist. Think you go outside[9] and when you come back you find 'state B, comparison 1' (o o o o o). At this moment 'state A, comparison 1' does not exist, it is in your memory. In other words, you are looking at one box, and it changes from "o o o o" to "o o o o o". Have in mind you are looking within the box limits, you are comparing the circles in the box at two states, state A (o o o o) and then state B (o o o o o). In this case, in order to find a difference, you need to remember the state A, which is no longer in front of you. You are comparing actual state B to past state A; while state B is in front of you, state A is in your memory. The system, circles within cell's table limits, has undergone disequilibrium, or has changed. The fact here is that 'state A, comparison 1', is not the same as 'state B, comparison 1'; this represents disequilibrium in our model of reality and represent change in our model for nature, which uses information.

The informational facts are clear; you had different structures within the same limits at consecutive (different) states. Comparing the facts, you have one more circle within limits at state B, or equally, you have one less circle within same limits at state A. The same will happen with the other two comparisons: 2 and 3; you find different structures at different states. There are many arguments about what could have happened within the limits. This exercise is not related to the concept of comparison or change; it is going farther to the use of informational ability, the process you are not aware you are using while you create those informational facts. How can you make a simple comparison without having informational ability? How can you create any part of these models without informational ability?

When we look at the cybernetics structure, sometimes there is change at one level and there is no change at other levels. The level of reference used is the inertial level, according to the structure – meta-system, system, and sub-system– already mentioned. Let us go

---

[9] The idea of going outside and coming back is to reflect an action. Actions are our notion of change and from there, time.

back to Figure 1B and compare 'state A, comparison 1' (o o o o) to 'state A, comparison 2' (o o oo); we can say that one circle has been transported near its neighbor. This creates a new system because of the *transportation* of one part of the system. Now, when we compare 'state A, comparison 1' (o o o o) to 'state B, comparison 2' (o o o ●), we can say that last circle on the right has been transformed. This creates a new system because of the *transformation* of one part of the system. We can pair these changes –transportation and transformation– to the system concept and say that the transportation took place through a sub-system call *link*, and the transformation took place through a sub-system call *component*. This can match structure and disequilibrium within our Figure 2. System components produce transformations and system links serve for transportations.

Those comparisons, systemic and cybernetic, have come from your informational ability –a process "transparent" to living beings. In this process you have use analysis and synthesis, as well as memory, without noticing it. You use informational ability to *synthesize* the elements in a set, creating two wholes. You use informational ability to *analyze*, and compare the sets, the two new wholes just created, finding distinctive characteristics. The comparison between the two wholes takes us to define properties for systems – components, links. The comparisons between states require memory and help us define properties for cybernetics (process) – transformation, transportation. Then, these basic comparisons could match structure and disequilibrium in our model for reality, to systems and cybernetics in the physical models.

After visiting the physical point of view, let us see how sentences present reality. The language unit, the sentence, is composed of subject and predicate. The subject is the material structure within the limits we create. The predicate is the change created by the subject, or a description of the structure of the subject. Material changes are only possible because disequilibrium exists. It is important to include the sentence concept because the language is an important tool in the humankind communication process.

We have compared two systems at same state, present time –one picture. Let me say that comparison is static. You can take the time you want in order to compare both systems. That is not the case with the cybernetics comparison. It takes two states; and those states need to be different enough for us to perceive them. Then, let me say the cybernetics (process) comparison is dynamic –one film. Now, we can present a table summarizing a model for reality, see Figure 2. One Model For Reality.

Reality	
Structure	**Disequilibrium**
System	Process
-Component	-Transformation
-Link	-Transportation
Static	Dynamic
Picture	Film
Subject (Noun)	Predicate (verb)

Figure 2. One Model For Reality

To capture the foundations of reality, we created the scientific model. But when we apply the scientific model we are observers and we simplify and complicate reality at the same occurrence. The observer takes to the experiment all its preconceptions. Fundamental preconceptions are time and length; they are informational facts created by comparisons, do not exist in reality. The observer attributes meaning to the comparison through his informational ability and also for the disequilibrium and the structure among particles. The preconceptions are useful at some levels of reality, like our level, but are complications at a smaller level, like the atomic level, where we are not used to attributing meaning. Also, time and length brings complication when we abstract things such as cycles. If there are cycles around us –day-night, birth-death, etc.– we conclude that the universe must follow the same pattern, having a beginning and ending; this is a complication. Why do we need to attribute a creation time or a size to the universe?

Reviewing the model for nature again, we find that the model for nature, beside life and death, has the observer's and actor's point of view. But the observer usually does not bear in mind that all living beings attribute/create meaning, including the living beings observed. When working in organic chemistry, cellular biology, etc., some very likely living beings such as enzymes, RNA polymerase, and ribosomes are seen as robots working in mechanistic ways. But the "functioning" of these parts of the living beings shows informational ability, i.e. they are behaving like living beings directing matter. Then, we should review the conventional model of nature having in mind that material structures with informational ability, which direct matter, are living beings.

Let us review meaning with the idea of mechanism and comparison in mind. In order to direct matter, let's say make a decision and chose a direction, you need to attribute/create meaning, and then make a comparison, and finally, make a decision over which direction to take; this is from the perspective of living matter. Now, you can attribute meaning and say that dead matter can direct matter, but in this case the dead matter does not have informational ability. Indeed, as we said, there are material particles in act. The act is structural disequilibrium, mechanical interactions, where the structure flows by inertia or action at distance. A river is an example. The water flows following different surface levels by combined actions among liquid water level, volume, gravity, environmental temperature, etc. and doing so carves the river path in its way.

To summarize the model for nature, the living beings we observe are elemental particle structures, aligning disequilibrium to maintain their living structures, creating an informational equilibrium. From multiple variables acting at the same moment, the living being acts in many different ways, creating patterns other living beings cannot attribute meaning to. The higher the informational ability level, the higher the possibility of having good models and better performance. Nature can also be seen as the cooperation among living beings to maintain life. We can see a living as a synthesis of elemental particles forming one special structure with the concept of fulfilling an informational principle, exist.

Objects

Objects in this book are always information, representations of matter, its absence or more abstract attributions/creations of meaning. Do not forget: matter creates beings. Living beings create information out of matter, creating a circular reference for matter: matter trials to matter. We can classify objects by the attribution or creation of meaning. First order objects can be called facts. We can split facts in two: material and informational facts. Second order objects come from the creation of meaning, are not reference to matter.

The direct attribution of meaning to structures or densities creates first order objects. Examples are any of the six table cells –system x, comparison y– in Figure 1A. These objects are based on what we see. This direct attribution of meaning gives us one point of view in informational ability, concreteness. They are material facts, concrete objects, first order objects. They are representations, meaning assigned to densities. We can see these first order objects as pictures.

When drawing distinctions between systems A-B at comparisons 1, 2, and 3, we end up with distinctions that differentiate them – quantity, quality, interactions. Those distinctions are called properties or characteristics, and they are objects too. These objects that are based on the distinctions we found, not on the matter itself, can be called informational facts. The informational facts require another point of view in informational ability; they require abstractness. These properties or characteristics that do not exist in matter and we living beings abstract them when comparing, lead us to later triangulate to other comparisons, and match other "properties of matter". These are not properties of matter, but abstractions we make from the structure or the disequilibrium. Since these abstractions are based on matter, we can use the word facts and call them informational facts.

Let us look at the triangulation or comparison of informational facts, and why we see them as matter's "own" characteristics. Using 'comparison 1, system A' (o o o o), we can compare it to a system's characteristics in previous distinctions; distinctions we have learned or created before –lines, leveling, distance, actions– and find some distinctions like:

a. The circles (elemental particles) form a line or are aligned.
b. The circles' line is horizontal.
c. The circles are distanced equally from the next in line.
d. The interactions among circles are given without elements of visible connection –there are no dashes among them.

The triangulation is made by first having the distinction; second, memorizing the distinction; third, when we see the new system structure, comparing it to the distinction. We see them as properties of matter since we are attributing meaning from our mind. We do not reflect on the fact that the comparison was made, the meaning attributed and memorized for later use.

As said, second order objects come from the creation of meaning; do not represent matter or any part of matter. Second order objects are abstractions, do not exist in matter. Abstract models exist for each leaving being; they are the way living beings compete for resources, creating meaning. The creation of meaning is responsible for translating electromagnetic waves to colors, mechanical waves to sounds, different molecules types to different smells, etc. We can see second order object responsible for the creation of models like reality, nature, philosophical, physical, etc. Models are also responsible for fear, rage, love, duty, order, reproduction, growth, maturation, and many other concepts, but those models required more informational ability. Later in Chapter 5, we will discuss some basic informational concepts. So, we have objects, objects within objects (models), and with them, we have objects' structures, like those structures of matter already presented in Chapter 1.

All languages are second order objects, and therefore models. There is no way to connect the symbols in the writing to reality without the triangulation already mentioned and the abstractions that lead us to a second order objects. So communication is a model created by living beings.

Using our informational ability, we need to exercise good criteria when attributing/creating meaning -information. We are living material structures -organisms-, from the first molecule giving direction to matter, to the greatest community on Earth. Each community creates codes that help its members to share living; this process is called communication. With communication in place, communities share all the material work required to maintain the direction of matter and preserve the living structure. Communities, to reach results −preserve ourselves− have two jobs. The first job is to select the direction the community needs to undergo. The second job is to take matter in the direction the community has decided. Both jobs need professionalism and respect. The professionalism goes two ways: from the living being to the community, producing what the community needs from its profession, and from the community to the professional, when the living being receives what he needs to live from other professionals. Here, the profession is not an academic title, it is a service performed on the best abilities of the professional. The respect requires that each living being has consistency between the job offered (what was said: information) and the job presented (what was done: matter) to the community.

I made an effort to describe matter without having to show personal inclination. But that is not possible. Languages are second-order objects, and they are accompanied by cultural elements. The central idea in this first part is that the particles of the universe have their own individual characteristics. Some individual characteristics are lost when these particles group together, acquiring new properties – the properties of the structure that they formed. The groups of particles have meaning for the beings that can attribute/create meaning to them, and are meaningless to those without informational ability. To exist is something concrete; that does not depend on any interpretation. The universe is here and always will be here, regardless of the meaning living beings give to it. To live is a process performed by complex material densities which requires informational ability.

Second Part:

Information, Trials To Matter.

"Knowledge is a viaticum, thinking is of primary necessity, the truth is a food like wheat." Victor Hugo

Information, from the practical point of view, is a collection of facts or data that has meaning. From a philosophical point of view, information is associated with the truth, referring to the idea that there are principles that create an absolute truth and that ultimately we can understand. For this book, information is the result of the informational ability. Informational ability is a process coming from the structure and functioning of living matter, this process allows the creation of organisms. Informational ability is a necessary condition for living beings, without it, there are no living beings. Informational ability is not a sufficient but a necessary condition for life. As we saw in the structure of living matter, living beings are structures of groups of elemental particles, in this sense, no different from structures of inert matter. But informational ability directs the process that maintains the structure of living beings, and in maintaining the structure, informational ability is maintained.

Before proceeding, let us define by the numbers –one, two and three– (1, 2 and 3) three concepts: one being, two objects, and three elements. Beings are made of matter. Objects are informational representations of the beings. The elements can be defined as one or the other, interchangeably –beings, objects or pure abstractions. To help us, let's see an example: one car being, two car objects and three car elements. The car is the being. The two cars are discriminated by two living beings (you reader and me writer) creating objects and three elements, one car being plus two car objects.

Let us imagine that we are a fundamental particle instead of a person. That is, try to imagine that you are a particle. We "see" ourselves the way we see the moon, one solid unit. There would be no internal movement and we would not think we would combine and recombine with other particles, no matter if we were part of gold, silver or diamonds; we would just be a particle that could decide anything. We would be part of the structure of the parts of matter participating in countless actions, running only by random effect under the basic properties of matter[10].

At the other extreme, let us imagine that we are the universe. We are all that exists, we are countless particles having stellar structures, and all structures are contained within you or me. You do not see anything because you are all that exists, there is nothing around. The galaxies move in groups, but this movement is random and under the emergent properties of specific groups of matter, stellar bodies. At some time a star explodes; at another, a black hole, a quasar is formed, etc. We are by ourselves following material properties.

Of course, these are the extremes that exist in matter: the universe and elemental particles. Between the extremes are very different structures, as we saw in the first part of this book, Matter. These structures operate by the basic properties of matter but when we take a group of particles that form a living being, this group of material processes includes a fundamental process that let

---

[10] Charge and inertia.

distinguishes living matter from inert matter: the informational ability. Such ability enables this structure to intervene in a specific direction. When this happens these structures are making or routing the operation in one direction that can be defined as one decision. For now, let us say that there is a sub-process, discrimination, which includes the notion of conscience and is the basis of the whole process of informational ability. Without discrimination you could not create concepts: there would be no information, although particles would continue to move.

We have talked about the structure of particles as the elements that we see. But remember that the particles are beings whose charge we cannot see. The charges of particles are also structured when the particles come together. This is something not quite mentioned, but charges of particles also combine to form a new charge, forming a new magnetic field. Then, the elemental particles with their charges, when combined, structure magnetic fields with emergent properties. Thus, we can say that the charges interact and their interaction forms an emerging magnetic field, which can also be considered a system. In summary, there are structures of particles and structures of charges of the particles; the two structures exist, and it can be said that if we consider them to be independent, as two systems, they exteract or if we take them as sub-systems, they interact, forming a system.

Some deny the particle's structure, saying that particles are too small and not worth considering. From there, they assume only magnetic fields, which are intangible[11], represent energy and discard the particles mass. We want to point out the coexistence, so in addition to the structure of particles, we must also bear in mind the structure of magnetic fields when we speak of beings, living or inert. As those familiar with radio frequency may know, altering magnetic fields create waves. The waves overlap each other,

---

[11] Intangibles that cannot be touch, even though can be perceive.

creating wave structures. Wave structures can be given by amplitude or frequency modulation. Recovering these waves requires use an inverse process to the one in their creation. For this you use a property or phenomenon: resonance. Resonance is a condition that occurs as a process, a dynamic equilibrium, where matter moves in harmony and creates disturbances. In the case of magnetic fields, those disturbances are useful to gather the generated radio waves; in the case of other physical systems, the disturbances are destructive. An example of useful applications of resonance is communication radio waves and laser light. In destructive effects, the most famous in my memory is the Tacoma Bridge in the U.S., where the wind destroyed a suspension bridge when it came in resonance with the structure. So charges that we do not see are also structured, and are part of the systems and processes that affect the structures of the particles.

As an introduction to the next chapter, let us say information is a state of the process of living matter which combines both the structure of the particles and the structure of the magnetic charges of the same matter, without which we lose sight of the existence of movement. In other words, we have a system –a being with an emergent property, information– which is the common essence of living matter and allows us to consider that being alive. Matter and information share the universe, as does the system-process model. Without matter there is no information, but there is still matter without information. This is the same as saying, without matter there is no movement, but there is still matter without movement.

Let's conceptually discuss what are information and the process of creating it, informational ability.

Chapter 4.

What is Information?

"...information is given and taken on a "need to know" basis. In other words, the bacteria prepare, send and accept the genetic message when the information is relevant to their existence." (Jacob & Shapira)

Information is the result of informational ability. Informational ability comes from a material disequilibrium that emerges in a material structure, organisms; which in turn is able to create informational structures, living being. This process includes sub-processes perception, discrimination, accrual, comparison, and judgment. While in the material disequilibrium, elemental particles and their properties create new systems; in the informational process only properties defined in the particular truth exist, information. The properties of the particular truth are defined by living being in the process of living. Keep in mind that a human being emerges from the process of a system that has trillions of cells, each one a living being. A eukaryotic cell, one living being, emerges from the process of a system that has a few prokaryotic cells, each a living being. Traditionally, we have three levels of living beings: prokaryotic, eukaryotic and multicellular; we can see it through the model –sub-system, system, and meta-system. Each set of living beings is a new living being with its own information. The informational process occurs throughout the structure, through each level of the structure that makes up a living being, and it is refined at each level.

Information can be conceived as giving shape to the structure of matter; but in this case, it is not about transforming the shaped matter, the being, but about attributing meaning to what is perceived, creating objects and models. Matter is required by both the being observed and the being observing; that is why we are presenting information as trials to matter. Matter is being judged by other matter, and the parameters are set by the matter which is judging, with its particular truth. The information has no pre-set parameters outside of the living being, so that is why the individual or collective responsibility rests with each one of us and in us all.

A key element in the informational process is comparing, and it can be stated that by using informational ability we are comparing. When distinguishing matter densities' change, compared density changes are seen as boundaries that define objects, no matter how fuzzy they are. So for information to exist, to create limits to "form" an object, differences are required, and the informational ability process builds comparing those differences.

The informational process starts, when the living being perceives shocks or waves from other particles or group of particles. These are received by the base of the living structure, let's say, the cells. Example: cells in an eye receive light rays; they are interpreted, and then will be communicated to other cells within the nervous system, which in turn, communicate with the brain and give final interpretation to the perceptions discriminated by eye cells. In this case, the final discrimination of the brain, an object, is the truth of the living being who is watching. The brain-mind is the highest informational point in the human structure, it has the last word. Those objects are accrued for reference. When resuming the process, if we have reference objects, we compare those discriminated against those accrued. Finally, we judge by deciding on one of multiple comparisons that we have at a time, and answer questions like: Is this the first time that this discrimination is before our eyes? Is it useful? Is it beautiful? Etc.

To Perceive

To perceive is to receive an action. The being that receives an action is the receiver. The actions among elemental particles can occur by collision of particles or by waves. The waves are alterations of magnetic fields. A group of actions form a process. The processes exist as actions of matter which create perturbations by action of their mass and/or charge. These perturbations have an effect on other matter that is in the course of the perturbation. The perturbation affects both inert matter and living matter. According to their intensity, perturbations can alter the structure of the matter that receives it or simply move or shake the structure. In the first case, the matter that receives a severe collision can break apart, be deformed, or if it is a living matter, lose its informational ability. In the second case, when the intensity is low, the perturbation will only create particles' motion without changing their structure. Until here, we can say that all matter, inert or alive, has received the perturbation. When perception is by collision, we are part of the action; other matter has impacted us or has hit us. When the perception is by waves, we perceive the waves from the perturbation; we have received no "direct" action, but action at distance. Let us say that we have received only a perception, like a spectator. There are waves whose intensity can destroy the physical structure; it is what happens with an explosion, as said, structure is altered.

Almost every word has different acceptions and each person chooses the desired interpretation. To better understand what we mean by perceiving and to create a reference to the informational process, let us discuss the hypothetical case of those whose mind goes blank.

Suppose that one morning, you open your eyes and everything is white. No shadows, everything is the same color -white. You think about it and wonder, "Am I looking at something white?" Then you look up and see white, look down and see white, left and right etc. Wherever you look is the same; everything looks white. You are literally in white. There is nothing before your eyes; it just appears white.

You can argue that many things have happened. Several volumes would be insufficient to list all of them. Then let us limit ourselves to some of them. As I just said, you opened your eyes and all was white. Let me add that you are twenty-two and were not born blind. That is, if you opened your eyes and saw white, it was not because you had always seen white. Nor was it that you were viewing a white sheet. It was not a white paper covering your eyes. It may seem you are watching a white something, because there are no shadows. Are you seeing a perfectly white thing? That is not the case, perfection does not exist. So you could think that you are dead. Since you have other senses, you think, "Let me try to hear conversations or other noises to verify that I am not dead." Then, you do not hear anything. So if you get "disconnected" from all your senses at the moment you open your eyes, it would mean no sense of smell, touch, etc. You would seriously be wondering if you are alive or dead. But let us insist you are alive. By the fact that you have no perception of your senses, you're not dead. You are isolated, a kind of kidnapping, not a material one, but an informational one. Of course, as you are without your senses, you will definitely die. How do you look for water if you do not see? Perhaps you could scream for water. But what happens when you get a response if you do not hear? They can bring water, but you cannot take it; you do not have senses, you do not know that they brought it. If they have brought it, you do not feel it in your mouth, because you have no sense of touch. They cannot take you anywhere until you ask them to. Suppose you are taken home, how do you know they are carrying you? You have no senses; no one knows where you are. You are like a stone by virtue of being completely cut off from your environment. You have your brain informational ability intact, but you are missing your afferent nervous system. Definitely, you are at the inertia of having been disconnected from the world, you are already disconnected from matter. You become matter incapable of executing decisions; you do not perceive your environment, and you will most likely die.

Becoming blank in this case was a purely informational issue. A rock does not go blank this way; a rock cannot think. It does not have informational ability. From the material point of view, the actions of matter occur through waves or by particle collisions or collisions between groups of particles according to their structure. We can say that matter perceives the actions of other matter. An example can be the moon. Have you looked at the moon and seen its craters? If the moon did not perceive the actions of meteorites, its surface would not crater. It would not have craters. If the moon did not perceive the Earth, it would not rotate around the Earth. Perhaps it would go around the sun, or it would just move straight ahead without being attracted to any stellar body. Then the moon perceives interactions with matter, just as any living being.

Another example of perceiving is that of a stone warmed up under the hot sun. If the stone does not perceive, the stone does not get warm, but the stone is heated by the sun, so the stone perceives the sun. What happens is that this perception has no meaning, does not generate information, and does not create an object within the stone.

Then, matter perturbates matter through its actions. Perturbations create a set of perceptions, not only on the subject that initiates them, but also on the subject that receives them. A perturbation is divided into action and reaction. The set of perturbations creates changes in the structures, and that is what we call movement. All existing processes are what we call the universe at work. In our model, to detect movement requires informational ability; otherwise how would we give meaning to different structures and their changes?

To Discriminate

To discriminate is to give limits to what is perceived. To discriminate is to define sections of what is perceived. Discriminating creates the images from what is perceived. A perception is a fact that we call data or event. In essence, to discriminate is the first step to create objects from what is perceived. In this step of the informational process, the living being is gathering actions from matter and turning them into objects. The structural conditions of living being let it "see" the matter. Depending on the informational ability, discrimination has more or less elements. For us humans with five senses, each sense presents its object. In any case, discrimination is created by the living being that perceives the action.

We perceive, then we discriminate what is perceived discovering imperfections; we find that there are differences. First, as material beings we perceive. Then, as living beings we discriminate one or more differences from what we have perceived. When we discriminate, we have a whole. The discrimination is possible because we have informational ability. When a living being has more informational ability, it has better discriminating abilities. In the same process, when we continue discriminating a whole, when defining new limits, we do not have a whole but rather parts. Now after we have parts, each can be defined as a separate object. When we further define parts as independent, we create objects, multiple wholes. Continuing this process, in theory we can get to the elemental particles, the basis of matter. Here is an exercise. Close your eyes and turn your head to your right; at the moment that you open your eyes, you look at one whole. It's everything you see when you just open your eyes, but then you begin discriminating. You review colors, continuity, distribution, and at the end of this process, you discover something. Let us say you were at a beachfront during sunset time. In this process, the initial image is everything. By following the colors, you discriminate particularly the sun, its intensity, etc. Then you see a particularly large block, which is the sea. When you continue discriminating, you see soft traces that are the clouds; with more discrimination you may see cumulus, nimbus, etc. The more discrimination being applied, the more objects catch your attention and so on, until you reach the physical limit of your eyes.

This leads us to understand that information is not out there waiting to be collected. It is not accrued outside of us; we created it internally. With or without the assistance of others, we create information by discriminating what is perceived. The creation of information begins with discrimination and arrives at the judgment of matter, which is the meaning we give to the perception received. That interpretation is our own, unique for every living being, and is created at the moment it is being discriminated. To understand this process a bit, look what Mira y Lopez says about fear:

> *"Let us make an imaginative effort and try to represent the origins of life on our planet. According to Haeckel's ideas, we can assume that the first living beings appeared in the plant kingdom, in the bottom of the sea, where environmental changes are relatively soft and slow, so luckily the conservation of any metabolic rate is easier; it's almost like, at a given time, ad hoc grouping of complex carbon molecules, the rings belonging to the series of organic chemistry, were created themselves, and the first protoplasmic micelles emerged, may be not yet specifically structured as stable forms, much less as individualisable macroscopically. Well, since then, in that primitive protoplasm, it is assumed that their micelles, upon receiving the impact of the new or sudden changes in the physicochemical environment (changes in osmotic pressure, electric charge, etc.) reveal a change in their metabolic rate, which is momentarily or permanently-engaged when the unevenness between the alterative capacity from the*

exterior and the resistant of its interior is inclined to favor the first (exciting or stimulating). And then they can occur in one colloidal precipitation process, more or less extensive, that is, a phase of 'gelification' that depending on being reversible or irreversible (according to the resilience on life) will determine a primitive state of colloidal 'shock' or protoplasmic 'death'. "

As explained by Mira y Lopez, changes in the environment perceived by the living being create a change in the functioning (rhythm) of that living being. In other words, the discriminating process of the living being creates change in its structure, a reaction that changes the living being itself. The change that occurs in the system environment creates changes in the structure of the particles or the structure of the magnetic fields. In this passage, the change represents fear, a living being is giving meaning to the perception of one change in the environment, and the action that occurs in the environment is discriminated as potentially harmful.

At the human level that is a high informational level, and using the sense of sight, let us analyze Figure 3. Picture To Discriminate. When we look at this picture for the first time, it is not easy to know what's there. What do those irregular lines on the concrete represent? This image is the path of one snail on a sidewalk. If you are not familiar with this, you will not reach a conclusion from the image. The other curiosity is that the path of the snail gets highlighted with the reflection of light, in this case, by sunlight, and is hard to see in the opposite direction.

Figure 3. Picture To Discriminate

For living beings, the easiest thing to detect is movement. To detect movement requires less effort, less informational ability. If we are looking at something before our eyes, we are seeking to identify the content of what is there; there will be colors that stand out more than others and attract our attention, but we have to attribute meaning, and the only way to do that is by discriminating. Giving limits to what we see and looking at our brain for some reference to what is before our eyes, we associate the colors, we follow continuous, contiguous points, etc. But in the case of motion, that which is moving distracts us and changes our general focus to what is moving. Thus, the discrimination –the informational process– is re-focused, and discrimination goes about what is moving.

The step of discrimination in itself becomes a process, because as we saw in the exercise, when we look we attach meaning. We separate the elements of what we see, analyzing; we organize the elements of what we see, synthesizing. Usually we do not have to think about this process; we do it unconsciously, discriminating by familiar association to what we have accrued in our brain. The process of discriminating is more difficult than bringing the already discriminated and accrued. To discriminate requires creating a whole. For this we must create boundaries, conceptualize, and so on. It's easier to check back to what we have in memory.

There is a second order level of discrimination. When we do not perceive, we can still discriminate; we discriminate on objects already created that are in our minds. We return images to present; we make mental representations and discriminations. This exercise brings us to present images that are not in the present, abstracting usually what struck us the most. This abstraction can lead to conclusions that do not exist in the general truth. They are ideas about ideas, among them, are emotions or logic. The operation of matter does not follow informational rules, we assign them to matter.

To Accrue

To accrue is to save what was discriminated, is to create memory. Accruing information requires material means, and the accrual also requires structure, an appropriate code. These materials means, as we have said, do not refer only to the mass of particles, but also to the charge of them. As we know, we do not see the gravitational interaction, but we see its effects, we are in constant attraction toward the center of the Earth. Then in the process of accruing what is already discriminated, there are influences from both particles' characteristics, mass and charge, and also the structure of the particles.

Here is an example. Suppose you think there is information in books. You would like to read Confucius. Let us say you cannot read Chinese and I happen to give you some writings of Confucius in Chinese. Can you, after receiving the book, start "collecting" the information in there? No. At the moment I give you the book, what it is in there is no information to you. You cannot discriminate

Chinese code. What you see are a few strokes, signs without meaning to you.

Let us analyze the example a bit. There is Confucian information in that book. Someone invented Chinese. Confucius learned it. Then, using the Chinese code, Confucius translated parts of his information by the use of abstraction, a part of informational ability, and codified the discriminations that he made, then materialized them in that book I gave you. Technically, Confucius accrued partial information in the book; he coded it. Now, you must first understand the code; that means discriminating meaningless strokes from Chinese language to learn Chinese. Then you can read (discriminate) the code to reverse what it represents. Reading it will be discriminating against and matching the code with the ideas you learned from your experience. As you do not have the same brain, nor have lived the same experiences as Confucius, you will give your own meaning to what Confucius wrote. If the text speaks of a fish, you imagine one that you have memorized, not the one Confucius saw. If he says he was watching the Yang River, you imagine a river that you have seen, not the one that Confucius saw. Even knowing the Yang River, you should be in the same place, with the same trees, at the same season and other countless factors in order to repeat Confucius' experience.

Reviewing this process, you will have a lot of work before you start giving meaning to the book's code. This is because you have not accrued, do not have memories, do not have Chinese code references, etc. After "learning" Chinese you will work to attribute meaning to what Confucius wrote, which should match your experiences, according to the information you have in your brain. Confucius wanted to communicate, and he partially got it through the book; the book is a channel that has no message for the living being that does not have the code, and whatever you visualize when you read the book are images of your prior experiences, not from the experiences of Confucius. The book will not have any of

Confucius' actual discriminatory experiences; the book only conveys concepts that help you in the process of creating your own information.

Do you still think that books contain information? Let us try to reach an agreement. Involving the next step in the informational process, to compare, let us say that books have signs which you are discriminating against. So you compare what was discriminated in the signs with your memory. You look for the sign presentation, and you decide what meaning these signs have for you; you pair those signs with other information you have accrued in your brain. In the end, at every moment of your reading, you refer to accrued information to interpret the signs in the book, and you attribute meaning to them. Your meaning is not information available to the writer, because simply, in physical terms, you are at the other end of the communication, you will judge with your informational ability.

To Compare

Beside discrimination is comparison. To discriminate is to give meaning to what we see; this creates objects. Comparing requires two elements. When comparing, the two elements may be two people, two objects or a combination. Here, to compare refers to distinguish one object discriminated from one object accrued. Technically, we compare objects because the beings are presented to the living being through the senses and through the memory (accrued). The comparison requires discrimination.

By comparing we are giving meaning to the present based on the meaning we gave to the past, which was accrued. We can extend meaning to the future based on probabilities. The mind represents our present; our brain simultaneously processes many actions, but we only have one of them present, which is currently displayed in the mind. To review this and other steps in the mind –what we perceive, what we discriminate, what we accrue and what we compare– look at what happens in one of examples of Mira y Lopez

(López, 1965). It is the case of a small dog in a violent situation, which could also be the case of some human beings (including myself).

"...our giant is one of the fastest and most savvy learners that are known. Take, for example, what happens to a dog, a few weeks old, when a man going in a car descends from it, shouts at it in a peculiar way and gives it a strong cane stroke on the back; for several days or weeks they will be bound as effective stimuli (i.e., the dog is conditioned) to determine its fear and flight reaction to all those elements which serve on the situation (constellation of events) that was painful. So, if it sees anyone only descend from any moving vehicle; perceives any shout similar to that preceding its pain; sees any individual with a stick, etc. it would be afraid. This has vastly increased the chances of suffering the clawing of fear without a real need.

... And finally, taking the case of animals that have an existential sense, it is clear that such fears – understandable but unjustified – increase suffering unnecessarily, in a blind attempt to avoid it. Because, in turn, each of them creates one hundred scares, and of this sort, will engender a kind of vicious circle that feeds up our giant, making it reach unusual proportions; those would come to invalidate us for any action, unless because in that degree of evolution have emerged from its own womb others, which ignoring its paternity, they will oppose it [fear] fiercely."

In this narrative, we can see that there was an action, the caning, and there were other elements around the scene of the event, which can be summarized as a constellation of elements, the truth of an event. At the time of the significant fact, the caning, the dog discriminated various objects: car, person, cane, shout, etc. The discrimination process was not very successful; the fact of the stroke was unclear, any of the objects of the event could have been the cause. The dog did not clearly discriminate what happened, did not understand the process, the comparisons lead him to the conclusion that the main fact could be repeated in the presence of at least one of the objects discriminated and now accrued.

In general, an analysis is a comparison. The comparison is an interpretation between objects, usually from different points of view. The dog looks to care for itself. But it is too young, and does not know who is who, or what is what. That is, who is the source: the car, the shout of the human, or the cane? All are objects associated as participants of the event. The dog lacks informational ability to understand the fact, and to know what really happened; the dog created its own truth out of the facts -objects. With its truth of the facts, the dog gives meaning to what may or may not be meaningful for us. The dog's analysis was poor; it has less experience and informational ability.

Let us talk about comparisons of second order, objects made of objects. We mention comparison between a perceived object and accrued object, but we can have comparison between two accrued objects. This comparison will lead to create second order objects as was mentioned in Chapter 3. Comparisons can lead us to say that two beings are equal, when in fact, what are equal are the objects, not the beings. The beings that are being discriminated can be defined by certain parameters. These parameters allow the living being creating the objects to compare them and say that the beings are equal, which is not true. In reality, common elements have been extracted, and by comparing those common elements that define the objects compared, we are comparing the objects, not the beings. An example: we say that two triangles are equal. We refer to the models we create, not the graphics that are on paper. The graphs on paper, easily abstracted, are different. To draw a triangle on paper, ink particles pass from one place to another, from the pen to paper. This makes us create a being at a time. We drew two triangles, which are two distinct beings. We agree on the abstraction, where the two triangles are equal objects representing beings, not on the fact, where the beings we create making lines on the paper are different. These two beings are supporting geometric ideas, "three lines form a triangle." Then we agreed to create equality where there is none, we do that through the use of discrimination and comparison.

Comparisons are made in any analytical process. Discrimination requires synthetical and analytical processes. All material facts and informational facts should be distinguished; facts result from informational sub-processes, which combined [sub-process] create an informational process. Material facts come from the elemental particles' densities; we can say we attribute meaning to those densities. From the particles themselves, material facts are data, come from discrimination. In our model of nature, we mentioned distinctions between structures; these distinctions –as information– do not exist for the densities themselves, and we called them models. These models [distinctions] are informational facts. They do not exist in our model for reality, are meaning creation. Classically, facts are bound to ones (1) and zeros (0); ones exist and zeros do not exist. We can bind ones and zeros to informational existence as informational facts.

To Judge

To judge is to assign meaning to a discrimination made. In other words, the informational process starts in the step of discriminating and ends in the step of judging. Then the "last" step in the informational process, where we get the result, where we finally create the information, is judging. The mind creates or recreates an object when it judges. When we are discriminating, we said, we synthesize. When we are comparing, we said, we analyze; but how many times should one review the discrimination/comparison process before judging? Each living being informational process determines which of the discriminations is most relevant for that living being, each of us. We are in charge of the informational process; we decide when to end any step, the process, or what is the result of it; we decide the meaning.

Let's look at an example referring to Figure 3, Picture To Discriminate. When asked, "What do you see in the picture?" Some observers will not give importance to the reflections. Suppose that you notice them. At the same time, you are separating the reflections from other objects such as the sidewalk, grass, and so on. That is an analysis. Then, if you ask, "What can these reflections be?" You are synthesizing, thinking about the reflections as a group. You can further analyze and find more than one reflection. Then you can count them, and the number of reflections is a synthesis. The process is discriminatory from the point of view of the whole, the question "What do you see in the picture?" The process is comparative when you relate elements. You will stay in the process as long as you want, and your final judgments may range from, "I do not know," to other sort of judgments, up to the "right answer."

Judgments are decisions. When judging, we are making a decision; we decide on the best of the discriminations. When synthesizing, the judgment is made by the integration of elements. When analyzing, the judgment is made by decomposing elements. Within the informational process we used both: analysis and synthesis. The objective of the living being judgments is to give meaning to the comparison of the current situation with others in memory. In the case of the dog, the small dog judges over which objects it must take care from: the car, the man or the cane, a couple of them or all of them.

There are times that, when deciding on actions between the informational and the material worlds, we create virtual reality. During a film presentation, our informational process allows us to "create" movement where none exists. We see people or other beings moving on a screen. But people are not really moving. What we see on the screen is a presentation of still photographs that, with the use of special mechanisms, seem as if something is happening; but what we are seeing doesn't exist; there is storage of facts, it is clear that we are "cheating" ourselves. The projector requires a minimum of twenty-five frames per second, displayed in a specific way, to give the impression of movement; this also means that the notion of information requires dynamics, comparison between present and past "realities".

The universe functions, works; it is at present. Motion is accepted as a fact but with no comparison between actual objects and accrued objects, there is no way to detect movement. Any informational ability requires references to discriminate, accrue, compare and finally judge movement.

Objectives are born with living beings. They are part of their particular truth; objectives have to do with what the living beings want for themselves and the beings around them: intentions. Objectives aim to meet needs. No inert being has objectives but human beings assign objectives to them. For example, someone whose objective is to build a better world creates artifacts to provide food more easily (sub-objective), but a user decides that there are other sub-objective for those artifacts. He turns them in weapons to dominate others. The initial idea of spears should have been to improve hunting, but the focus shifted to a different one when we started to use them against ourselves. Another example: stars emit light. Our star, the sun, does too. You can say that our sun has an objective: to bring energy for us. The fact is there, the sun is there; you attribute the objective.

When there is more informational ability there is the possibility to include more points of view in the informational process. In this case, we are talking about "animals that have an existential emotion", and living beings that have desires/motives (López, 1965). These are purely informational concepts, where the needs are secondary; they appeal to ideas of second order: emotions. We are saying emotions are pure informational concepts, because there is no physical need for them, but they increase our informational ability. Then, emotions integrate the informational ability of some living beings, making them better equipped if they do not lose other points of view, like physical needs point of view. We will talk more about consciousness later, which relate to emotions, and motives.

Judging is the final stage of the informational process. We must bear in mind the simultaneity of the actions of matter when we think about the large number of cells that make up our brain. Thus, while part of the brain is perceiving, another part is discriminating, another part is comparing, another part is accruing, and another part is judging; we need to bear that the brain is working in every step of the informational process at all times, making it more difficult to understand the informational process, and to understand that information is our creation as living beings.

Chapter 5.

Where Is The Information?

"As conscious beings, we exist only in response to other things, and we cannot know ourselves at all without knowing them." Harry G. Frankfurt

We have already said information is the result of the informational ability of living beings. This informational ability allows us to create objects that put together with our expectations or goals, create a general model, which is our philosophy. In a more graphic way, we can say that the information is a system, a photo, and the informational ability is a process that creates the movie. That movie is also information; the life that every living being creates is its nature.

Let us review the notion of objects are information. The objects are created by the brain directly from the information it receives from our senses. Let us note that these "first" hand objects depend on the configuration of our senses. Discriminating a vibration makes it an object in the same way as discriminating a symphony is also an object. Similarly, the discrimination of a point is an object like the discrimination of a movie. The initial elements of the discrimination are not within our direct reach, they depend on the characteristics of the cells that make up each of the senses. The afferent nervous system cells carry the discrimination made by the senses from what was perceived to the brain, and there, the brain cells combine the discrimination obtaining an object from the perception. The object is live in an essential emergent brain process, the mind. One object in the mind is created by the structure and disequilibrium of many cells at a time. Those cells have a common function, living; and common informational processes, not from a specific cell but by the group of cells. As mentioned, cells form specialized groups, colonies, responsible for the functions of the different organs; in turn, various organs are combined into a higher structure, an individual body. Both the living structure of the cell and the structure of the individual have their "own" information; the information of the individual emerges from cell structures. The objects in our mind are an informational creation, which results from the information that nerve cells have carried to the brain, and the brain process: the mind.

Then, objects are an informational creation from living beings. But what do we call more developed objects? That is, what do we call objects such as triangles? They are objects created when we extract few features from objects, or let us say objects based on informational parameters of objects. We saw the outlines of the triangles on paper, coming from informational parameters, three lines and three angles, where we create another object, the triangle. Let us call this object type, model.

In themselves, models are objects, but those can be called "second" hand objects. Here's another example: "Car." The word car is representing millions of objects in a second hand model: English. In itself, it does not define a particular object, but through English, it is part of a transportation concept. This characteristic that distances the being -car- from the model or code -car- makes writing one of the most special human-created models.

To create models you need adequate information ability, and you need to be able to abstract characteristics or parameters from two beings and make them equal, accordingly. Thus, every living being is grouping objects and relating them to create models, with which it confronts the world of matter; in other words, every living being discriminates and creates boundaries that define objects, then attaches properties to them and creates models. One special example is the model for science, based on the scientific method, with which we greatly differentiate ourselves from other animals. But that model has limitations like any other; when we separate those beings under scientific study, they lose emergent properties, and therefore the application of the scientific method does not allow us to study processes like informational ability. We require new methods to overcome the uncertainty mentioned by Heisenberg; the scientific method has limits in its material application.

This chapter presents a model with which we can make sense of, or at least give context to, an informational whole within living beings. You can call it system, brain-mind, brain, mind for animals, state-government, state, government for societies. The model shows one informational unit, from which you get parameters to use your informational ability. That unit commands the whole of each living being, the truth for each living. The unit emerges from many sub-units or parts. In all cases, informational ability/information is the driving force of life. Examples of this start from many complex organic molecules forming prokaryotic cells, and continue up to animals-plants forming ecosystems. The scientific model gets short explaining the information gathering for each living group, i.e. all our neurons, that leads us to one unit; but that unit is real and is where information belongs: the living matter which is the final reference to what happens as general truth.

Particular Truth, The Life Philosophy

Every living being in the structure of life has its own information. Each living being is trapped in it; your information is your reality; what you have processed with your informational ability is accrued within you. The reality of every living being can be called its particular truth. The word truth is used because you have created those objects that exist in you, and you used them as true. By combining these objects, you create models, and those models which have cost you effort are true; they are particularly true to you. This truth is particular to you; that is why we call this model: particular truth.

Our particular truth changes when we learn because we incorporate other objects in our model, we reaffirm them again, or we change some objects for other objects. When we learn, our model for nature is transformed; we have changed our particular truth.

Reviewing the history of mankind, we find that ancients have created models according to their particular truth, which, in turn, is consistent with the informational ability of the moment and in accordance with technological advances that are common to these groups through history. This concept is what Thomas Kuhn called scientific paradigms.

The truth is a second order concept, a model. It is a model which has had many martyrs. First order concepts are process from the facts, what happens. These facts, as we said, are perceived and discriminated by the senses, which are cells, and carried to the brain, the place where your particular truth (the reader's truth) is. In the brain by an unknown process, you discriminate, accrue, compare and judge in "simultaneous" actions. When the brain does this process, it is based on your particular truth, which can be seen as your philosophy of life, one global reference to all your actions. When we say global reference, it is not a single reference; it is a set of references for each point of view. An example: in politics, you can be part of one political party or another; in family, another point of view is that you can either think that children must blindly obey parents or that children and parents should have criteria to determine when to obey and when not. Then, when you reach an opinion or decision you have used your particular truth because no matter what others may think, if you think you know the truth, you know the truth.

A practical reference to the particular truth, a scientific one, is the notion of color. Colors only exist in each of us, for the blind they do not exist. We can say that when we are looking at a flower, say red, you perceive the same color that I perceive. Look at daltonic people who confuse red and green as an extreme. What happens is that when we go to school, we are "synchronized". This synchronization helps with language and with colors at the same time. I'm not saying that there are important differences, but there are some. It is common that when color differences are not so pronounced, we discuss what the tone color in paintings or photographs is. We can say that the daltonic person has impaired color vision and they see what is green as red and vice versa. This confirms that the colors we see in nature do not really have a preset chart, and that as there is a swap of colors in the daltonic, the brain may represent colors (electromagnetic waves) differently from that which we are familiar with. As reference in this sense, visual discrimination of perception of electromagnetic waves, some animals can perceive what we do not perceive.

With the term particular truth, we are talking about all the information we have as reference and which in turn we use in the creation of new information. The interpretation of the relevant environment, the "outside world", is particular to each of us. This is commonly called subjectivity; but in reality, what we think is objective because it is created by us when we judge. So the personal model with which we look at the world is the particular truth. When we judge, we create informational filters for the actions of others with our particular truth. In other words, what we learn when raised in a family becomes a "permanent" reference while we are alive. So a healthy particular truth enables us to align our views in one direction, to live better according to what we believe. Our particular truth becomes a particular philosophy.

That is why philosophy, and more specifically, our philosophy of life, is as important as the material of which we are made up. All our points of view branch out from our life philosophy. To mention some points of view let me refer to several authors:

Stephen Covey, in *First Things First*, says we have material and informational needs: to live, to love, to learn and to leave a legacy. There, he also says that we have four gifts (informational ability) to help us in the process: self-knowledge, consciousness, independent will, and creative imagination.

Katherine Benziger, in *Thriving In Mind*, tells us that we have four brains which see differently, i.e. there are four ways to see nature: structure, rhythm, logic, and concept; maximization is achieved with their integrated use.

Emilio Mira y Lopez, in *The Four Giants Of The Soul*, speaks of fear, anger, love, and duty. He stresses that a sense of duty is a young giant that has its entire splendor in leaders.

Later in the book's third section, the section about life, we will talk more about the topic of control, but it is now important to quote a point of view of personal freedom (the person as a whole in society) and freedom of the individual in society (the person as a part of society). This quote challenges external control with internal control, or what is called locus of control. The site where control or governance lies is fundamental in understanding how to create the particular truth in order to form a healthy living community.

> *"Men are qualified for civil liberty in exact proportion to their disposition to put moral chains upon their own appetites; in proportion as their love to justice is above their rapacity; in proportion as their soundness and sobriety of understanding is above their vanity and presumption; in proportion as they are more disposed to listen to the counsels of the wise and good, in preference to the flattery of knaves. Society cannot exist, unless a controlling power upon will and appetite be placed somewhere; and the less of it there is within, the more there must be without. It is ordained in the eternal constitution of*

things, that men of intemperate minds cannot be free. Their passions forge their fetters. " Edmund Burke, 1791.

Freedom and control go hand in hand; they are set by our particular truth. Our philosophy of life or our particular truth is a reference that we create in our lives, our own informational reference for decision-making. This reference takes us to obey our parents, or other authorities, without any doubt, or precede what some call criteria, and others, individual ethics. What we have highlighted with the particular truth is that living beings, while having the same genetic basis, perceive the world differently because we have different informational and material structures. The informational structure is the particular truth, the material structure is given by genetics, but the informational ability includes both and it changes when we learn.

When we make difficult decisions above everyone else, including parents, spouse, children, etcetera, it is because we think we are right. And within our particular truth, which represents the life philosophy of each of us, we are right. But the important thing is to understand the consequences of that particular truth. Let us understand that people around us are support, they have other models of reality, then respecting and confronting their particular truth with ours, we will achieve synergy and with that, a superior outcome to our individual outcome; and with that comparison everyone will have a better model of the general truth. There are people who think that matter, the outcome, the final system, is the most important part of life. Others think that the information, the building, the process, is the most important part. In this book, we think that both impact our lives; the building processes are as important as the results. Both matter and information impact our informational ability, one with material limits and other with informational limits. Your material limits create your body; your informational limits create your philosophy; without one or the other you are not you.

It is important to note here that this whole informational process takes place within the natural characteristics of matter. There is no such thing as supernatural beings; there is ignorance. As exposed by Michel Largo, "There has been an eternal quest for the divine," a quest to find the truth and specially the general truth.

So each being has its own particular truth, with which it leads its own life by the "path" that each one decides; that is, the particular truth limits the informational ability and also limits the objectives we define to achieve. They are all within our particular truth. A truly healthy particular truth creates an aligned life, where what is possible and impossible for each one of us is learned, without optimistic or pessimistic prejudices.

Conventional Truth, Culture & Science

You have a particular truth; I have another truth. We cannot have the same particular truth; the truth is particularly unique because of all the elements it contains. But there are elements that the two of us share. There are elements in your particular truth that match elements in my particular truth. These elements of your particular truth that match mine become common to you and me. They become conventional truth; we are "sharing" models or part of them.

These common elements help us communicate concepts; when we communicate, we create a greater living being. This new living being, created by ties of conventional truth, is a new system and it has new limits. If we form a team, you and I are part of a larger system; you and I are sub-systems. Larger systems can make larger processes. In economics there is the concept of specialization which allows greater productivity, and in turn, more wealth. Specialization cannot exist without conventional truth.

The elements of the conventional truth come from second-order objects: scientific model, political model, natural model. When making the first drawings in caves, the human being began an artistic process to reproduce objects "taking them back" to beings. The forerunners of today's writing are hieroglyphs, an intermediate step between the drawing and the abstraction of today's writing. There are other second-order objects which came with this process, like beauty, freedom, etc.; they were born in a practice-theory process, which leads to the expansion of language and abstraction. In this process humans created a basic language model, theory, which then increased in the process of using the model, practice. Every culture is full of conventional truth. So much, that we could say that the conventional truth is culture; it is made by groups of living beings that share models. Note: It is conventional truth for groups of psychologists that 90% of our behavior or performance comes from the conventional truth received or learned before age seven.

While you read these lines you are sharing a conventional truth, one language, English. We agree on this code, a model that is agreed on in advance between you and me, because it was born prior to us as a creation of conventionalism. To create conventionalism takes hundreds or thousands of years; this has happened for languages. The process of higher language began with the creation of multicellular animals, which in turn requires other less abstract language. It went through hand-signals, sign language, the sounds that we hear in some animals today, and then the pioneers in the written language, humans; we created strokes in the cave, and as we said, through hieroglyphics, we came to the written word. Today, thanks to the written word, we develop video, supplementing media and showing the actions of matter in a virtual way improving our communication tools. We exchange code, which is information inside us, using our informational ability. Let us insist then, that information does not exist outside of the living, it is the essence of the life to create it. Perhaps that is why it is so difficult to communicate, the entire process is internal. Covey suggests that the easiest way to communicate is to understand the particular truth of the living being with which we want to communicate, so we can refer our ideas in terms of the other, in terms of what the other person understands (Covey, 1990)[12].

Algorithms, recipes, and science are all procedures coming from nature. The point of view to emphasize is the abstraction required to form first or second order conventional objects-models. Science helps us find out how matter works. Scientists ask questions and then develop recipes, technically called algorithms, to separate pieces of matter, add other pieces of matter, run appliances and mix, to then perceive, discriminate, accrue, compare and judge the answers. When other humans can repeat these recipes and get the same results, we are witnessing a scientific process. The scientific method documentation allow other scientists to review that the mix of matter, according to defined algorithm, always behave on the same way, or we can say that the experimenter always sees the same process results. This method, used for many years, has allowed us to recognize some behavior of matter and has opened many possibilities. These opportunities are growing every day and will be greater to the extent that we may find more conventional truth without ruining or killing our particular truth.

Let us see a Russell passage which emphasizes justice within conventional truth and freedom within particular truth:

> *"Broadly speaking, we have distinguished two main purposes of social activities: on the one hand, security and justice require centralized governmental control, which must extend to the creation of a world government if it is to be effective. Progress, on the contrary, requires the utmost scope for personal initiative that is compatible with social order.*

---

[12] See the fifth habit in the book *The Seven Habits of Covey*: "Seek first to understand, then to be understood."

In cultural matters, diversity is a condition of progress. Bodies that have a certain independence of the State, such as universities and learned societies, have great value in this respect. It is deplorable to see, as in present-day Russia, men of science compelled to subscribe to obscurantist nonsense at the behest of scientifically ignorant politicians who are able and willing to enforce their ridiculous decisions by the use of economic and police power. Such pitiful spectacles can only be prevented by limiting the activities of politicians to the sphere in which they may be supposed competent. They should not presume to decide what is good music, or good biology, or good philosophy. I should not wish such matters to be decided in this country by the personal taste of any Prime Minister, past, present, or future, even if, by good luck, his taste were impeccable.

I come now to the question of personal ethics, as opposed to the question of social and political institutions. No man is wholly free, and no man is wholly a slave. To the extent to which a man has freedom, He needs a personal morality to guide his conduct. There are some who would say that a man need only obey the accepted moral code of his community. But I do not think any student of anthropology could be content with this answer. Such practices as cannibalism, human sacrifice, and head hunting have died out as a result of moral protests against conventional moral opinion. If a man seriously desires to live the best life that is open to him, he must learn to be critical of the tribal customs and tribal beliefs that are generally accepted among his neighbors." (Russell, 1949)

We have talked about the conventional truth at the human being level. These ideas must be applied equally to cells. Cells also have emergent properties and manage information, except we are just beginning to understand the functioning of cells. We have not yet cracked down the codes they use to communicate, although it is a fact that they have to communicate to create multicellular living beings. Contemporary anatomy and physiology textbooks present natural processes mechanically, without the notion of any meaning or interpretation by the livings involved. But life emerges from matter, when that matter creates meaning and is able to redirect matter. We only notice cells' actions, as when we see the actions of other living beings or the actions of the stars. Cells, like us, use physical means to communicate, such particles and magnetic fields that generate electricity and motion. But we do not have all of the codes to understand what is said among them. We do not see their information; we only see the flow of matter even as electrical impulses. It is as simple as our inability to understand a language strange to us. Information has been historically presented outside of living being, with external existence; but the information is inside the living beings –particular truth–, as result of its informational ability, and among living beings –conventional truth– to give life to a greater living being.

Speaking of the cell's conventional truth, the university work *"Meaning-Based Intelligence"* (Jacob & Shapira, 2004), speaks clearly about the cooperative performance of colonies of bacteria, from 10^9 to 10^{13}, which are grouped together and communicate by various means, producing a collective work without losing their individual character. The cells are changing their particular truth, and adding conventional truth. We can say that they learn when sharing with other cells, acquiring conventional truth like a group. This shows us how single-celled organisms can share information, creating a conventional truth, while keeping individual characteristics, maintaining their particular truth. Extending this concept to material structure, we found that with the help of the conventional truth, new living beings are created, multicellular beings that also have a new informational case, around which they operate a common objective; the "new" living being is a multicellular organism. In this new living being, the bond strength is given by informational effects, communication.

What is explained above is seen more easily in structures of living beings at our level, in herds. Every drove develops a conventional truth that ends up distinguishing some drove from others, i.e. each group has conventional truths that characterize them. In essence, we have the same structural concept as seen emerging from matter; individually some properties are displayed, particular truth, and collectively other properties are displayed, conventional truth. These informational models are replicated in insects like ants or bees and in large herds of animals like zebras or deer, groups of birds, etc. This conventional truth is seen as "egalitarian" behavior, and has no informational meaning for us; we define it as behavioral instinct but it is social behavior as "reasonable" as ours. As an example of droves developing their own behaviors, we can take groups of people. According to anthropological analysis, all humans have a common ancestor, the hominids group. If this is possible, how do we explain different languages? Languages are conventional truth; social groups evolved in isolation, developing different language; leaders exerting authority, or different informational abilities among humans, are among many reasons why this may have occurred.

In conclusion, the conventional truth allows integration of living beings. That integration creates groups that can be seen as one living being sharing concepts. Communication occurs as a result of creating conventional truth -models- among living beings at the same level.

General Truth, Matter

We can say that the truth of truths -general truth- is the world we try to model, matter. No matter what we perceive or discriminate, either individually or collectively, the basis of all that exists is those elemental particles, they are the indisputable truth. Elemental particles are not seen, but the actions of the particles and their properties that enable all to be structured are seen. With each structure come different results; the stars, water, and living beings are made of particles. Living beings are part of those structures of particles; the big difference is that our work or operation is guided by information, by individual goals, without escape from the particle properties and their structures; we are a material structure capable of giving meaning to matter. Matter is the general truth when it comes to facts. Let us see an example. A group of doctors can discuss particular truths when asked, how did that person die? What was the cause that led him there, the trigger? What was the time of the last breath? But the fact, the general and unique truth that they all agree on, is that there is a dead body.

If the general truth is given by facts, then these facts are rooted in the structure of matter. As mentioned in the first section, matter is there; it is elemental particles with mass and charge. We, the living beings, are one structure that discriminates from what we perceive; the actions of matter –their shocks or their waves. Let's look at an example. Sound, the sound waves reach the eyes, the eyes perceive the mechanical vibrations, but give no interpretation to them. The same happens when your doctor shine's light in your ear to look at the tympanum. The tympanum receives the light, but you cannot hear it. The eye is composed of cells, including those that discriminate electromagnetic waves, and those cells send the discrimination to the brain, where the object is created. The ear also has cells that discriminate mechanical vibrations and sends information to the brain where the object is created.

In physical science, there are models that aim to predict the actions of matter. These models are based on first-order objects, and taken further to second-order objects by physicists, through the scientific method and mathematics. This is a whole informational process in our brains supported by its environment –the general truth. When having a model, physicists perform the reverse process, and with experimentation verify again and again what was discriminated. With these nature-reality model cycles, we intend to model the general truth and predict what will happen; this is the same theory-practice process. But that modeling is far from the general truth –it is just enough to be conventional truth– especially when working with large sets of elemental particles. The vast majority of actions among particles are filtered. We turn part of these actions into information. Hence, we do not capture reality, only a "simple" model of what reality is –our model of nature. An example of this filter is the short range of the electromagnetic spectrum we capture.

It should be noted that the facts in the general truth, have been and shall be the beginning of all things; facts are the actions of matter. Someone told me I'm being too radical with my model of information. But the model is stated accordingly: matter has mass and charge and between the two properties, particles have the option to create and recreate the most diverse shapes and processes. As parts, the particles create wholes with internal dynamics that are affected by external dynamics. These wholes eventually acquire a basic informational ability, the lowest form of life. This minimal form of life is difficult to distinguish from inert matter; especially if we are still looking this minimal amount of life with a mechanical approach. We need to go further with the informational approach, but I have faith that we will understand the concept and identify the transition from the living to the inert or vice versa. At our level, this minimal living being appears to be a machine. The distance between the stimulus and response is so physical, so positive, that we do not see an information concept; its actions seem like simple physical and chemical reactions that ultimately, at the atomic level, reflect as structure changes. Changing from one structure to another can be interpreted in many ways. At the very base, these minimal living beings are matter which attribute meaning to other matter that surrounds them with a very basic and simple interpretation: its usefulness. The matter in its environment helps it or does not help it live. That is why, despite the difficulties, for many communities are the best way to live. For this reason, some forms of life are structured in more complex ways, with a high degree of informational ability, which lets them use more information. This leads to multilevel living beings –they usually see beyond short-term usefulness. Similarly, today in politics, we continue to negotiate the best structure or order to organize humankind.

The general truth is represented by matter. It is in us because we are matter, and it is outside of us because otherwise we would not perceive anything. The general truth teaches living beings the limits of their capabilities. In the same way in which a physicist explores the world, in the nature-reality model cycle, every living being explores the world. Trial and error is such an intuitive process that we do not notice it, but it is the confrontation between general truth and particular truth. Reality is the general truth, continual shocks and disturbances transforming the structures formed by elemental particles. In the general truth there is no concept of the universe, no concept of perfection, no concept of necessity, or desire; elemental particles only exist in actions without discrimination.

To show the truth of matter we can refer to the fact that despite what we think, when we break the material support to the process of life, the information creation process ends, as the process of creating light ends by removing its material support.

Let us look at some informational concepts.

Chapter 6.

Ten Basic Informational Concepts.

"A good world needs knowledge, kindliness, and courage; it does not need a regretful hankering after the past or a fettering of the free intelligence by the words uttered long ago by ignorant men."
Bertrand Russell

We have argued that information is the result of a process, informational ability, which only takes place within living being; it is an internal process; it is within living beings, then without the existence of living beings there is no information. Material actions continue; the universe will work without us. These informational concepts we are going to see are for reference; they are discriminations of first-level or second-level; they are objects or models that exit within living beings.

Let us remember, complexity comes from the number of elements – beings and objects– and the number of events among them, because the principles governing matter are simple. Look at the case of the Euclidean Point, an axiom of geometry with the same name which has allowed mathematics' development. The Euclidean Point does not exist in the general truth, only in the conventional truth; it is a base object, an axiom in Euclidean geometry. Before Euclid, geometry was practical, used by experience. Some call it the method of exhaustion, others, trial and error, to work over and over again to obtain the results. The idea of a point allows us to abstract all matter and then create an object, an indisputable starting point, and once it is accepted as an axiom, it is conventional truth. If a point has an infinitesimal measure, as some say, then it exists in the material world; it is a very small sphere. Therefore, Euclid made a great abstraction and created something that really does not exist in the general truth, and the same has happened with other informational concepts that are more or less clear. With the help of practical geometry, shapes and the patterns of nature, and first-class objects Euclid developed a point model, a second-level object. Thereafter, theoretical geometry allowed new developments of abstract thinking and scientific improvements, dialectical development between theory and practice.

Then, let us look at some basic informational concepts that seem to be an important first-hand challenge for many who say that "it is logical that living beings are so, because they manage information," but are not aware that logic starts with the acceptance of clauses truth. This is not technically true; the first step of living beings is to create information and the second step is to use it or manage it to create more information making decisions. We will start this tour in Perfection, which represents homogeneity in the material world. We will continue with Points of View, that is, where or under which basis you make judgments. Logic, which are particular truth rules that the general truth does not "know," will be followed by Structure, the concept that lets us create our own universe, objects, and lets some diminish the small particles and their ceaseless interaction. Then we will analyze Cycles, which only exist when accompanied by a reference. Equilibrium, the non-existent made real. Then Measurement is logic applied to object comparison to create equality. This will be followed by Numbers and their daughter Math, model of models. Time, almost a pure child of information, is supported by the concept of cycles. Finally, we will discuss The Information Trap, matter's killer or big egos' creator.

1. Perfection, Homogeneity.

Perfection is a concept; it does not exist in the general truth. Perfection is hypothetical because to get to the material world, something needs to be imperfect, and to get at least some equilibrium we need to pay a price planning, executing, and controlling. Moreover, let us say, it does not exist in the conventional truth, because we all have our own idea of perfection; therefore it exists in our particular truth.

Let us look at an example. Have you noticed that what for you is perfect, for your counterpart who is on the other side of the desk, it is not. I have a friend who is very important to me. He has climbed the corporate ladder and complained about how the government wanted to restrict his commercial freedom. He was speaking as a customer at one side of the desk. When asked whether he restricts his employees, he answered without a hesitation: "Of course, how else would I achieve results?" at the other side of the desk. Then it is perfect to restrict others, but it is not perfect that we be restricted. However, the differences in the way of thinking about perfection, if properly managed, will create expertise from every point of view, and through the combined synergy, will create the best results.

From the material point of view, perfection can be associated with homogeneity. Every concept has points of view, and perfection is not the exception. Perfection in itself means nothing. Perhaps it most closely approximates to having no information or being unable to have it. Perfection from the point of view of interactions would mean that there are no errors –those all actions comply with the same pattern and do not come out of that pattern. From the standpoint of beings, structured matter, it would mean that the whole is homogeneous and has no impurity. From the point of view of society, some say, perfection would be like equality, no one has more than anyone else, everyone dresses the same, and everyone looks the same. But in matter, it is given as just the opposite, disequilibrium, different shapes, processes with "errors", etc. Then from the material point of view, there is no perfection.

From the information point of view, in pure abstraction, ideas of second order, we could speak of perfection. Everything works in the world of ideas. You dream of castles with many rooms that do not need cleaning because they stay clean if you say so. Also, there is no heat, no cold, and your castle is at such a comfortable temperature that you do not even think about it. We may, as in the fairy tales, kings' stories, or any perfect story, spend many pages talking about perfection.

Perfection within objects, let us say perfection as an adjective, can also be conceived as the being that is defined by specifications and behaves according to them. That is, by the concept that we created for it; in terms of accomplishing objectives, what we modeled for it is what it does. Example: if you're talking about a person, this person is perfect as a person if it behaves as such. A politician, according to some people I know, is defined as "someone who speaks well, does some of what he promises, and steals" (I must say that there are exceptions). A tiger, if it has stripes, hunts and eats meat, then is a tiger. In those cases, the person, the politician, and the tiger came perfect person, politician and tiger when behave as defined. Please accept some of my oversimplifications here and elsewhere; it is a way to create information. In short, they meet the definition of what they are, and then they are perfect. Then there is perfection of adhering to the specifications, attained by following the qualifications we attribute to beings.

But, to behave as defined is not necessarily perfection. Where is the definition of the tasks that need to be performed by the tiger? In the sense we mentioned, the tiger will be perfect for a group that shares what the tiger should do in order to be a tiger and there already would be several definitions of perfection for the tiger. This is what happens with experts. They are grouped according to their particular truths and then begin to discuss a particular idea. From a more abstract concept, for the tiger to be perfect, in addition to completing the tiger tasks, the tiger should perform the tasks perfectly. That is, the tiger should not get sick, should keep the same weight and all hunting expeditions must end with a big beautiful prey, etc.

As we said, homogeneity does not exist in the material world. Let us use our system-process model, Figure 2. If matter is made from particles, we cannot talk of homogeneity; there is discontinuity, which is not a clear idea of perfection. But let us accept that there is homogeneity still having particles, which represent discontinuity. Let us look at the structure. Suppose that the particles are equal, but what about the distances among them? If they are not equal, as we know, there is no homogeneity. Suppose, in order to continue, that the particles are equal and that the distances among them are also equal. What about the dynamics among them? Would they all move at the same time, or will they not move? If we assume that they all move together, i.e. at the same time, synchronized in the same direction, it's like not moving at all. How do you know if they move when they all move at the same speed? We need a reference, but which one? All particles move as a block. In order to have homogeneity within matter, all the properties should be equal, which would not let us discriminate the particles, because everything would be the same. In that hypothetical case, we would not exist.

Some days we are more optimistic and other days we are more pessimistic, there is no homogeneity. Informational ability is tested at any moment. Informational ability includes these features, ups and downs, because in general truth there is no perfection or homogeneity. While understanding that perfection is a concept developed by humans, we should use the concept to improve ourselves, to better discriminate, and to build a better world by improving concepts such as faith, courage and justice. The idea of perfection transform us, the human beings, to perfectible beings, beings guided by the idea of perfection.

2. Points of View

Points of view can be associated with matter or information. By associating the point of view to matter, we refer to the position where we look at the being observed. An example is when we look at a person. We can look at the person from the front, from the back or from one of its profiles.

By associating the point of view to information, we refer to the basis for the informational process we have. We use informational points of view called paradigms, which are filters. Suppose we are seeing two people, let us say a man and a woman; we can say they are a couple, one point of view; or they are two people, other point of view. A couple is a whole. Two people are two wholes. In the first point of view using synthesis we see a system. In the second point of view using analysis we look at the individuals; they are two parts. Here the informational processes have –synthesis and analysis– which are two information points of view. These concepts have occupied the psychologists who argued that the differences arise from the brain physiology or from the brain structure. One model we have heard of for many years talks about brain structure, and claims that the left and right hemisphere process the same event in two different ways, as two points of view. Look at it in the words of Covey:

> "A great deal of research has been conducted for decades on what has come to be called brain dominance theory. The findings basically indicate that each hemisphere of the brain –left and right– tends to specialize in and preside over different functions, process different kind of information, and deal with different kinds of problems.
> Essentially, the left hemisphere is the more logical/verbal one and the right hemisphere the more intuitive, creative

one. The left deals with words, the right deals with pictures; the left with parts and specifics, the right with wholes and the relationship between parts. The left deals with analysis, which means to break apart; the right with synthesis, which means to put together. The left deals with sequential thinking; the right with simultaneous and holistic thinking. The left is time bound; the right is time free.

Although people use both sides of the brain, one side or the other generally tends to be dominant in each individual. Of course, the ideal would be to cultivate and develop the ability to have good crossover between both sides of the brain so that a person could first sense what the situation called for and then use the appropriate tool to deal with it. But people tend to stay in the 'comfort zone' of their dominant hemisphere and process every situation according to either a right or left brain preference.

In the words of Abraham Maslow, 'He that is good with a hammer tends to think everything is a nail. '" (Covey, 1990)

Matter, as we have said, just is or exists. The material point of view, when something happens, is a fact or general truth. The informational point of view is the creation of information in every living being, which is particular to it. The brain hemispheres processing a fact in different ways represent multiple points of view from the same fact, two in this case, synthesis and analysis. Multiple informational points of view represent multiple meanings. Having multiple points of view creates relativism. Then relativism is due to uniqueness of the informational ability in each individual when judging the same fact.

An example: it seems pointless to say that the one who receives an action does not have the same point of view as the one who executes it; it is obvious! But how to distinguish between an accident and a well disguised malicious intent? The fact does not change; the result of it is. Premeditation is not easy to measure, but the one receiving the action has a point of view and the one who executed it has another point of view. When we judge if something was intentional any witness has an interpretation of the facts and with them, their judgment. This leads us to believe that lawsuits in the public podium can be a matter of economic power, not justice.

Let's look at another example. If you regularly travel on a bicycle and purchase a car, does your world become bigger or smaller? Everything depends on your point of view. If you analyze in terms of how you use the bike, that is, moving in a two-mile area, the distance you travel will not change with the purchase of your car. Your world is probably getting smaller for that same distance, you travel faster, and you save time. Now, changing the point of view; you bought the car because you want go farther, and when traveling farther your world, includes more elements. You will see more variety; your world has become larger.

3. Logic

Logic is particular truth. Logic is a model which uses structures and process –rules and algorithms– to help the informational ability. In the scientific models, conventional logic is applied. An example of the logic used by scientists is the one created by Aristotle. Let us remember that models are used in science, and they are second order discriminations.

Let us look at the following proposition. If A = B and B = C, then A = C. Here is a model using different objects and assigning properties that do not correspond to reality.

1. There are no two equal beings; the general truth is unique and tangible. How can we say that A = B? By performing various abstractions. One abstraction is that A represents an object, that B represents another object, and that C yet another. The "=" equal sign is another object or abstraction.

2. We assign properties that confuse individuals with concrete minds. "Are you telling to me that "A = B" (A equals B)? Are you a moron? Where do you get that A equals B? Are you not able to compare and see the difference?"

In material terms, let's say we are looking at two hydrogen atoms. Because they are atoms of the same chemical element, by logic (rules) they are equal. Here we are entering chemistry; the hydrogen atoms are "equal" under the protonic model where an equal number of protons and electrons mean the atom is in balancing load. It is known that the electrons are moving around the nucleus randomly, according to energy and other factors. If electrons are moving and are not in the same position, assuming that their other features are equal, the two hydrogen atoms are not equal.

Logic goes through many informational filters. Logic does not reflect reality; it helps us create models which we defined as second order discrimination. Logic is information that only exists in some living beings, especially humans. As an exercise, think of a being which is equal to another being.

Through somebody's "logic" or the logic of a particular group, it is agreed upon that there are equal beings, that things are the same. But that logic does not reflect the world; it is based on conventionalisms. Besides filters of the senses, characteristics or properties of objects are being filtered, and thereby, we compare the discriminated models, and there, yes, we could say that A = B. Given that A is referring to a hydrogen atom and B is referring to another hydrogen atom, which under the protonic model are equal.

When objects are simple or there are few concepts, we may find characteristics within each system using the informational point of view that allows us to apply logic. If there are more objects, more structures, or more interactions with more advanced informational concepts or fuzzy logic like: kind (hateful), good (bad), useful (useless), terrible (friendly), beautiful (ugly), large (small), far (near), much (few), wide (narrow), the logic loses soundness; it makes no sense by the complexity of the concepts or the hassle of creating conventional truth from the particular truth of participants.

Living beings have more affinity with probabilities than with logic. One example is global warming, the change in Earth's temperature that is reducing the ice areas, poles, snow picks, etc., and will increase sea level. It is real, but it is not the most imminent danger for humanity, considering what a super-volcano, a large meteor or another quick but unforeseen event could do. However, the low probability of the event withdraws it from the "logic" of individuals, while the frequent media coverage of another attracts their "logic." Logically, Goliath had to kill David. Logically, the more intelligent from the logical point of view, high IQ, should be in charge, but it is usually the one with more emotional intelligence. Logic is a set of rules that helps organize models, but it does not ensure any result, nor does the concept of probability. Both help; while one is based on abstractions, logic, the other is based on real facts, on the count of events.

To close the concept of logic as a set of rules that seek to simplify reality, we can look at discussions among people. Many people argue, accusing one another of having no logic in their approach, how come? Finally, to leave an exercise bearing simplifications and the concept of human logic, think: what is more logical, that there be men and women or that we were all hermaphrodites?

4. Structure.

Structure is one of the most basic informational concepts. Structure is given by the discrimination of matter, more specifically to a subset of it. We create structures when we observe the world and how it works. Structure "becomes" part of the properties that we assign to matter, the shape that takes a set of particles, which we define within us as one whole or as one subset of particles from the whole, which we call part of the whole.

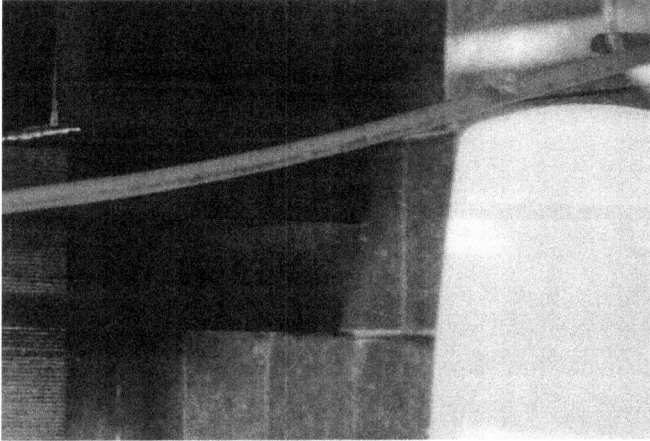

Figure 4. What Do You See Here?

The structures of matter are like clouds. Remember the old game of looking at the clouds and trying to see figures? See in Figure 4. What Do You See Here? The same thing happens here, when we create information, we use it latter to create more information. Then we work to structure our information, and give our particular truth our own structure. From that informational structure, you have a "clear" idea of when your family comes before your job and vice versa. Another example is in business. You receive information from the finance department or perhaps the operations research department. You use the informational structure of your particular truth and try to give meaning to it. Will it be that the quality of your service/product is not very good? Will it be maintenance problems, maybe the new product from the competition? You are trying to discriminate. You use a structure of elements, comparing with structures of other periods that you have accrued in your memory and eventually you get to the root of the problem using the informational structure of your ideas.

In developing the idea of structure, we are developing the idea of a system because we create groups of particles that serve as a reference to understand other group of particles. The structures we create are based on models that, as we have said, are abstractions which make sense; they are logical for some and not logical for others. Just remember the structures mentioned in the first part of the book.

How do we define whether a structure is in chaos or in order? Order is a basic idea for the living, aesthetic. The processes of life require certain way or there can be no life. In the first case, the system of life is an aesthetic issue, a preference for the concept that things should stand with symmetry or asymmetry. Everyone, implicitly or explicitly, defines certain aesthetic rules. Keeping them is order; not following them is disorder, which can be extended to or seen as chaos. In the second case, the processes of life require a certain order; prioritizing actions that combine informational and material actions is a must. The search for food requires discriminating and then material actions. In developing the idea of order, by extrapolation, the concept that there is a universal order comes to mind. The beings –living and nonliving– are routed, sorted, "working" together for something "useful". There is a direction among living beings toward a thing, goal, etc. Someone may say that order has nothing to do with utility, but with what we discussed, the process of life requires being consistent. From the point of view of your particular truth, you have your own idea of order and your order is what you decide. Also, what is right for you is right for you, regardless of others' order. But in the execution of projects, you will clearly understand that there is an order to build physical structures; without it, you cannot execute your plans. Differently from the structure of information, we give the order that we want to ideas, but that "order" is judged by others as order or chaos. In conclusion, in the informational world, everyone is correct regarding order, but in the material world, the order it is in the result, the final test of a "good" material structure or order is the preservation of life.

5. Cycles.

Have you noticed that you have never been in the same place in your life? Let us explain it. A place in space can be your home; it usually does not move from its building site when we take the Earth as a reference. You can say that there are cycles for you, from your home, which does not move. Each day is marked by the sunrise and each evening is marked by the sunset and so on. Is that not the definition of a cycle? From memory, you have been aware of this cycle: day and night. The same thing happens if you live in places where the Earth has defined seasons. You see that there are other cycles, spring-summer-autumn-winter, and the idea extends to the cycles of life, born-live-die, so many processes are treated as cycles. I said "with reference to the Earth." This is a restriction, – informational filter– on the model of general truth. In the general truth, the Earth is not the center of the universe, as many people had held for centuries. As we said in The Stellar Structure, we are in a galaxy. The sun moves around the galaxy, which takes nearly two hundred million years per cycle. The Earth moves around the sun, a cycle which we call a year. Adding the fact of galactic motion in the universe, you have never been in the same place in your life –that is a general truth. Your logic does not match this analysis. Your logic tells you that your house has been in the same place whole life, because your home does not move away from you, it is not moving at any moment, and you have been repeatedly in your house.

Then, having a reference, in this case the Earth is the basis for defining cycles. When we use other references, your house moves around the sun, the sun around the galaxy and the galaxy away from the center of explosion of the Big Bang (a concept with which I disagree). In the general truth, there are no cycles, but they are created by us as models that help us build conventional truth; those models help us interpret the world. Then the creation of cycles requires defining a point of reference and with this benchmark we start seeing some "cycles" and destroying others.

It is worth clarifying the notion of a cycle as information, not as movement of matter. Cycling is the movement of a being who goes through some stages and returns to the starting point. When you left the house in the morning, went to work and came back, an informational cycle was performed within the conventional truth. You abstract or ignore certain features of the whole to which they belong and then create cycles for yourself, in your particular truth. Let's look at another example. You say the school learning cycle – first years of school, elementary, middle school, high school and then college. If you have present that you do not reach the same place because each school segment is advancing you, it is not a cycle, it is a journey. Now, in the dialectical process, learning-applying can also be seen as a learning cycle; you learn the mathematical foundations and then apply them, but next time you are reinforcing then with more advanced mathematical knowledge. So it is not a clear cycle, either.

Perhaps you're familiar with the second law of thermodynamics, the law of entropy. I mention this because the law of entropy speaks of irreversible processes. If the processes are irreversible, they support the idea that there are no cycles in the general truth. If you are familiar with heat pumps, you can say that in an air-conditioning system, there are cycles: "The Freon is recycled over and over again and creates true cycles." While this recirculation is statistically not all particles move at the same time, and it is taking the Earth as a reference; in addition, the cooling system is drawing heat from a cool site that you want to remove it from, to the environment. Clearly, it is not a cycle, although some of that heat comes back through the walls.

Then, cycles are models, informational concepts, which require having points of view, and at least a good reference. The idea of cycles helps us understand the processes through which systems pass, but they are not part of the general truth.

6. Equilibrium

General equilibrium does not exist in matter. General equilibrium leads to the controversial idea of perfection. Where everything is perfect there are no errors. When everything is still, you cannot make mistakes. Everything is in balance, nothing moves and nothing crashes or changes. If nothing moves, the only possible error is that something starts moving. Then we can say that equilibrium is an abstraction of disequilibrium, an extrapolation. This extrapolation may lead us to say there can be at least one balance in the general truth, the universe is in a dynamic equilibrium; otherwise we would not exist. But the dynamic equilibrium in this case is the result of the disequilibrium of the elemental particles that can structure and restructure.

Perhaps your idea of balance is in one dimension, you think in a plane. But a flying plane is not in equilibrium. If it were in general equilibrium, it would not move. The plane is in "equilibrium" in two directions, vertical and horizontal, but in the travel direction the plane is completely disequilibrated with respect to the Earth. It moves in a direction, toward the destiny site, the target. This disequilibrium has focus on the plane and its reference is the Earth.

When a helicopter is in a "single" place, it is closer to equilibrium in three dimensions with reference to the Earth, but it is disequilibrating the surrounding air to stay at that single place. In this case the helicopter has no internal equilibrium, its parts are moving. In this sense, we will find idea of equilibrium filtering points of view.

Using the analysis point of view, which we use at the structure of the parts of matter, each time we agree on an external equilibrium we can see internal disequilibrium, until we reach atoms and find they are in disequilibrium. Now, let us look the opposite point of view, synthesis. If atoms were in equilibrium, we would not be alive, there would be no chemical reactions, electricity would not exist because there were no electron motions, all would be dark, and there wouldn't be any photons. So, from the base of matter there is disequilibrium creating changes, and living beings create the concept of movement comparing present to past.

In other informational point of view, to learn is disequilibrium. Learning requires disequilibrating the particular truth and the material system that gives origin to the information we have. Logically, the farther outside of our area of knowledge we go, the more we learn, but in reality, the farther we go, the more risk we face without any guarantee of learning more. Going too far and losing equilibrium can be fatal. Let us visualize this process of change, informational disequilibrium, by drawing three circles of different sizes, Figure 5 - Learning Rings. The inner circle is a comfort zone. The middle circle is an area of challenge. The last circle is an area of pain or passion. In terms of learning, this tells us that we learn the most when we take the greatest challenges, which are more demanding, but in turn, they carry more risks. In the comfort zone, there is virtually no risk, we are not stressed, but at the same time there is little learning. The intermediate circle is more challenging than the first, but less beneficial than the last (Colvin, 2008). In short, each of us must decide how much disequilibrium to give to our lives in order to learn. However, we must measure the risk conscientiously, because sometimes we risk losing everything when taking suicidal risks that are outside the areas of "normal" learning.

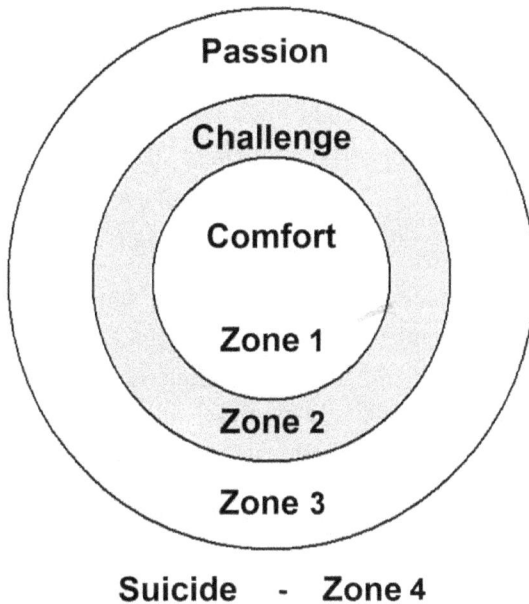

Figure 5. Learning Rings.

The incessant disequilibrium of particles is reflected in structure change. Each change can be seen as a set of events. There are at least three events: the trigger event or cause that initiates the change, the intermediate event, which is the transition or rearrangement, and the final event, which is appreciated with the new structure, i.e. with the new system in place. We, living beings, track, record, and classify these changes using our informational ability. But we overlook informational ability elements like fear, love, rage and duty, which present biases to the set of events that are tracked, recorded, and classified.

Then we can talk about an informational equilibrium, where by making use of information, we track events and give meaning to repeated actions as equilibrium, overlooking that any flow is disequilibrium.

7. Measurement

Measurement, in its most basic concept, is a process by which we compare objects. Comparing objects helps create models that allow us to create conventional truth, up to create equality. Measurement uses the logic of the transitive law that we have already seen, if A = B and B = C then A = C, or the triangulation process seen in Chapter 3

Let us look at distance measurement, length. A meter is an arbitrary length, a ten-millionth of a quadrant of an Earth meridian; this means nothing until we have something concrete to reference. Therefore, we have measured the meridian, a task that took years at the time, and built a standard bar of platinum that theoretically measures a $1x10^{-7}$ of a meridian, at zero degrees Celsius. When we say a being is one meter in length, we are comparing that being against the size of the Earth. For the measurement to take place, we get a replica of the standard meter and compare our being of interest, place them next to each other and compare and decide whether or not the being at hand measures the ten-millionth of a quadrant of an Earth meridian. By observing this, we understand that the replica of the standard meter is in the middle of the transitive law. If A=1m and 1m=B then object A is equal object B.

By increasing the number of points of view and by increasing the number of parties involved in the process of measuring, we come to other measurements or other comparisons that become more complex. For example, measuring the length of a piece of flat substantially elongated wood with defined borders, See Figure 6 - Table And Mass, using a meter ruler, is an easy measurement. That is what we do at school when we have to repeat laboratory tests. But let's say we are asked to measure the length of a lump of coal, which has an indefinite shape, clearly the form of a rock, see Figure 6. Table And Mass. The piece of coal, that we have intuitively clear, is one structure not easily measured in any way by a meter. Another tool is required to measure and we would have to measure several times to know its length, the longest distance between two ends of the lump.

If we want to measure the density, we need to know mass' weight and volume, and the case becomes more complex. Going farther, more conceptual measurements can be done i.e. movement: Will the mass of coal and piece of wood fall with the same acceleration? Energy: How can we compare the heat performance of wood and coal? In any case measurement requires careful selection of parameters and the tools to conduct the measurement. If we cannot reach an agreement through communication, the measurement only makes sense for the living being that performs the act of measuring. In this sense, we get to endless discussions when we deal with mental or physical tasks, which task is harder to plan or to build? And this also gets people to say that the task they performed is harder than ours.

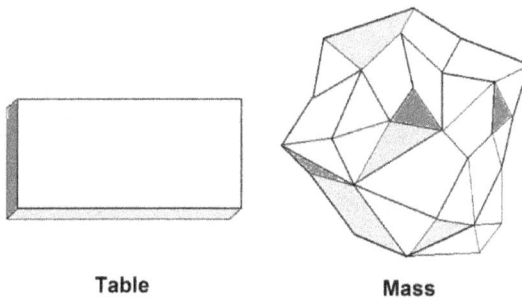

Table   Mass

Figure 6. Table And Mass.

The natural measurement given by comparison, leads us to judgments such as: two things are equal or two things are not equal.

But we need to base that equality on the same point of view. Examples: color, one is red and the other is green; appearance, one shines and the other does not; opacity, one is transparent the other is opaque; size, one is bigger than the other; speed, one is faster than the other, etc. In fuzzy logic, we talk about "littles", a little more or a little less. The littles are difficult to interpret, and with human development we have created measurement tools that support conventional truth. That was the idea behind the meter, to provide fairness in transactions between individuals; it emerged within the French Revolution seeking to improve trade (Alonso & Finn, 1992).

When we measure, we are pointing out characteristics of the objects we have created, overlooking the informational ability, which make to compare, creation of information and support for the conventional truth.

8. Numbers and Mathematics

Numbers are one of the most prominent concepts in information. The unit is the abstraction of all the properties of an object, turning it into a whole, one unit. This is also equivalent to the concept of a system. You define the parts that interact and see the system like a unit. Numbers have led a growing conceptualization process. First, natural numbers, then zero, negative numbers, then decimal numbers or fractions, real numbers, finally imaginary numbers, which are the creation of complex second order objects. Mathematics was born from that abstraction; it is based on numbers. Mathematics can be seen as a process, the processing of numbers. Mathematics allows the summarization of plenty of work in to a single figure, a number, or in the worst case, a few of them. An example is the measure of business activity. All company activities in a period of time are summarized in a number, profit. Also, in the company's balance sheet or cash flow few figures will summarize all the activities of many humans.

Numbers are invented by humans in an important process of abstraction. Let us remember that one of the most emphasized aspects in the mathematics classroom was: "You can only add things that are the same." Inventing numbers required that we initially compare objects and abstract qualities; we made them the "same" under a set of observed characteristics. In other words, we modeled reality and created numbers. Over time, that became "obvious." The abstraction of characteristics to define that two elements are equal is difficult, but once we identify that pears have a shape, color and characteristic flavor, we say that they are alike and different from the oranges; the equality becomes "evident". We add pears and oranges apart. But the initial concept was not easy. Then numbers are a significant informational advancement, another basic informational concept that is attached to mathematics.

We could also come up with numbers in other ways, such as trying to understand reality or controlling it. Example: one shepherd judges that his flock of sheep fluctuates, his pile of sheep changes size. He sees that some leave, others are trapped by wolves, and more surprisingly, they multiply; he decides to clearly point out what happens. He makes an abstraction and believes that in his cave, which serves as shelter, drawing a line per each sheep can be useful for this purpose. That is, a line will represent a sheep; he has gone through the entire process of discrimination and comparison, and it seems that now he has an informational need. He will not eat numbers, they will not help against the cold winter, physically they are useless, but as information, they are a real asset. He does not know he has fifteen sheep, he does not differentiate them that much or, if you prefer, he is a very successful shepherd and perhaps has forty five sheep. Anyway, the point is the informational need, and he solves it with strokes, which are numbers. This is easily said, but this process, as simple as it is for us today, could take generations, trial and error, different versions, and finally we began a process of numbers and mathematics, which is still evolving today. More and more abstract concepts are developing but we find the same basic informational process.

Mathematics is a quantifier language, not a qualifier language. When we say one hundred, we do not say anything. To be useful it requires us to define the limits and designate the objects of reference. Once we define the limits, we have a base object, which gives meaning to that one hundred: one hundred particles, one hundred groups of particles with the same form, one hundred apples. We know that in physics there are dimensionless numbers, but they are in a rather abstract context. Numbers require qualification; they require us to define the units. However, let us use dimensionless numbers and create an equation for life.

The elements in this equation are matter, information and matter's interactions. Planning (P) is thinking with an end in mind, using information to construct a model with a stated objective. Planning defines a process and/or a product using different presentation techniques. Executing (E) is working the plan and transforming material structures following a model-objective. Executing built a process and/or product. Controlling (C) is ensuring that the work follows the plan, measuring the transformations that are performed according to the established model-objective. The plan, the execution, and the control components are often based on characteristics according to the objective in mind. Life is the sum of planning and executing under an act of measuring the difference between the plan and the implementation –coordination/control. Keep in mind that the control should be given at each moment, creating symbiosis between planning, executing and control. Then a formula for living is:

Formula: To live $= (P+E)/C$

Planning and controlling have the informational elements which do not have preset properties and are the result of the philosophy of life, the use of information. Executing has material elements; it is limited by the properties of matter. If the plan was created based on general truth, the implementation will build what was planned. The new form that matter takes will be that modeled. Only living beings can plan. Plans distance the action of a matter on us, perception, from our reaction to that action, execution. Action-reaction is typical on inert matter –no plan involved. The control is parallel to plan and execution, because at each moment, we ask ourselves whether what we do is right, and we corroborate it with nature. The control in living beings is permanent. Certainly you have not noticed how many times you have inhaled in the last minute, or how many times your heart has pulsed. This is a control of another part of your nervous system, while you take care of other issues of your life, the external affairs.

As in the previous case of the formula for life, mathematics creates models that, as they are created, are applied to physics, already made discriminations. Physics comes from first-order information; mathematics was born as second-order object. Mathematics helps explain many physical phenomena, but does not qualify them. For example, the solid state is not defined by any mathematical number; it is defined by the feeling of stiffness of a structure. Technically, there are ways of "measuring" this stiffness. Some of these measurements are referred to as hardness testing. Some of the names of the measurement process and its scales are Vickers, Brain, Shore, Brinell, etc.

For those versed in vector algebra and the physical concept of work, the vector dot product of force times distance equals work done on a body, or equally is the energy required to move it. The mathematical function is: $W = F.S$, which represents the physical concept quite well. But according to this definition, you do not perform any work when loading a large suitcase on your shoulder walking on a flat surface. The truth is that it costs a lot of effort to perform that movement and it's exhausting. Most of us consider it a real job. So have in mind that the world is not created under any mathematical model; we apply mathematics to better understand and predict the world.

9. Time

Time is a pure object, like the Euclidean Point or the unit, which gives rise to mathematics. Time comes from the discrimination of changes in the structure of matter and the ability to accrue information. Time allows us to compare processes and within the conventional truth, allows the synchronization of actions among us.

We have talked about structures of matter and said that the change from one structure to another is a transition, one process. In our model, the initial structure is a system A, and the final structure is a system B. Let's propose that the system A is two vertical dots, ":". And the system B is two horizontal dots,"..". Between structure A":" and B "..", there is a transition, which we call process. How long did it take for you to go from structure A to B? Let's say you took one second. One second, in this example, is comparing a fraction of the Earth's rotation with the action of reading among the vertical and horizontal points. So the basis in the notion of time is also comparison, which in turn requires informational ability.

Having a reference is a prerequisite for creating information. References are only possible if there are differences. When speaking of perfection, we saw that unless there are differences, there is no way of referencing. If everything moves simultaneously, there is no reference to detect motion. Similarly, if everything is white and there is no difference in the white tone, there is no reference to detect any being. If you do not have informational ability, you can not refer to anything, you do not have the ability to discriminate whether the structure has changed or not. So when we reference change and we want to measure which one is faster, we need conventional truth. Up to now, we agree on what is a second, and so, by saying 100 meters (we already agree on meters) in 10 seconds we are taking references and making them common among us. When we want to compare with a runner 100 m/10 sec, we do not need to be with the runner to decide that we are faster than him; we use the concept of measurement and tools developed for this purpose.

Let us look from another point of view to discuss the informational concept of time. Let's say you have been diagnosed with a memory disease called Timeillness, a disease that allows you to recognize what you see while it is in front of you, but once it is removed from your sight, you do not remember what you saw. There comes a friend with some pictures. Your friend shows you the first picture. You recognize what you see, it looks very nice, a circle. Then your friend shows you the second picture, a hexagon. Finally, another photo shows a square.

When asked about the circle, you do not remember. What about the hexagon? You do not remember either; you just recognize the square that you have in front of your eyes. You can only speak of the square, as it is what you are seeing. When repeating the procedure, in an attempt to understand the situation, the same thing happens. You just remember the square that you have in front of eyes. Your friend does not give up; your friend decides to change the order, beginning with the square and ending on the circle. This time, you just remember the circle. What happens? You discriminate what you have in front of you. You only discriminate the picture you have in front of your eyes; you're in the present. Having no memory, there is no notion of the past. We humans have created the concept of time without noticing the use of memory in the process.

All actions of the universe are completed at the present. There is simultaneity of actions; none occurs in the past, nor does the universe plan them for the future. The continuing disequilibrium created by the actions of matter and limited by the properties of it is "extended" in the present. When living beings perceive the actions, they do it in the present. Of course, the discrimination of those perceptions is received by the brain after they have passed, by issue, not only of geography but by the symbiosis of material and informational processes.

Let us look at how geography affects the reception of events. Each set of events is different for every living being because of the geography or the place where the living being perceives the events. An example that proves this is that when we receive sunlight it has been traveling a little over eight minutes. But when we receive the light of a star it has been traveling for years. If we get both at a specific time, which is the time of the event, the time to receive the light from both celestial bodies, the perception of events happens in the present –here and now. However, the actions that gave rise to the light, about eight minutes earlier, and the other a few years before, are not happening at the present time. The light perception is a fact, it happens in the present when there is discrimination. But when does the event take place? When does the cones and rods discriminate it? Or when does the brain discriminate it? It takes some time to process, between the eyes and the brain and before the mind becomes aware of it. This is particular to each of us. It is the moment when we become aware of the event. Without informational ability, there is no concept creation, there is no information.

Let us look at another case. You perceive and discriminate a structure, say a landscape where there is a bird. You accrue the bird; it stands out in that landscape. In our model, from the One Model For Nature chapter, we said that there is dynamics when one part of the structure changes position. When the bird flies, you perceive and discriminate again, but this time you compare with the accrued information and detect the bird in another position, detecting changes in the structure. After the judgment, you conclude that the bird flew. You then discriminate using your informational ability and detect the change in position, with the help of informational process, to compare. This can be considered as objects of first order. You are comparing the here and now object, with the object accrued in your brain, giving meaning to the difference brings you the notion of time.

Looking at the idea of coordination from the informational point of view is only possible with the help of the concept of time. We humans have a variable perception of the concept of time. That is, we can spend hours in activities that we like without thinking about changing them. In these cases, it seems that time does not elapse. Whereas when performing an activity we do not like, let's say, waiting for someone without a distractions, any minute we wait seems eternal. Also it is very interesting, that for example, when using a computer, which increases the productivity or saves us a lot of work, we despair when it takes a little longer.

Returning to the idea of coordination, how could we coordinate activities without something like the idea of time? We could say that events like the sunset or sunrise would help us, but how to coordinate through rotation of the Earth or translations of the same? Calendars are a rather ingenious way of associating cycles to the creation of the idea of time, and there are calendars dating back to ancient times. In the early history of mankind, we worked with the projection of the sun. Buildings were constructed such that, at a certain time of the year, they allowed the builders to have a signal; the sun's rays passing through a hole, and reaching a demarcated point, indicated to start the planting season. Then, with further elaboration, with the help of numbers, we began conception or invention and the use of the calendar and the sun clock. And then, with other technological developments, not long ago, humans developed the mechanical clock and time was bound as a property of nature. Even in today's physics textbooks it is shown as part of the general truth.

When you remember, you do not return to the past; you use your informational ability. You can not physically return to that moment when you fell down, you can just remember it. Let us look at the other side, the side of the future. You think of something; you believe it should be done; you make your plan and the next day or if you prefer, after the sun rises, you carry out your plans. You did not travel to the future when you made your plans. You used your informational ability at all its glory, you planned-created something you wanted to do, prepared it, and always in the here and now, transformed the matter using information from the plan. You made it; the execution was an event at a time, event by event, all of them at present time and then finally you completed it; you finished a project in the present. You have completed all your activities in the present, not jumped to the next day; you slept, but did not travel to the future. All our activities are carried out in the present. Note that we speak of time as a measure, as a way of comparing movement.

Think about it, time travel does not exist. There is only walking, traveling by land, by water or by air, changing relative position, and if we can solve a couple of little material and informational problems, space travel will be as current as flying today.

Consider the theory of relativity. Let us remember an example about two different times. Is it possible for time to "run" at two speeds, or is time dependent on the reference? The example I refer to is that, if you board a spaceship and it exits through Space at the speed of light, on a trip which according to you is one light year long, when you return to Earth, according to an observer on it, one hundred years have passed. In other words, each observer has its "own" time and for the one who traveled at the speed of light, it took a year, but for the one who was on Earth one hundred years has passed. This can also be seen on the "contrary", that is, you want to save some time traveling and travel to space 365/100 days, or 3.65 days, at the speed of light on that spaceship. On Earth it has been a year. This example exists in the world of ideas, information, because we do not have the technology to try traveling at the speed of light. Let us create a giant camera that takes super photos fast enough to capture spaceships that travel at the speed of light. This giant camera keeps track of any spacecraft.

When the ship departs the first photo is taken and so on every 3.65 earthling calendar days. This gives us one hundred photos for the year. Since the camera is so special, where is the ship located at every photo? In each of them, the same time has gone by on Earth and in Space. Why must the metabolism of the individual who runs the spaceship slow down? The truth is that there is no time difference; at the end of the day, it is an idea. Reaching these speeds has technical inconveniences, the forces used to maintain the speed are enormous, no doubt, there are important technical issues, but the idea or concept is not altered. The second-order object, the model of time in this case, does not exist in reality. Time is an informational concept only altered by the living being; its idea of time does not change the general truth. Traveling at the speed of light does not make us younger. Comparing this idea with the notion of a triangle, a triangle does not change as long as the idea is that a triangle has three sides, three angles. Ultimately, time is a human creation and you cannot alter the concept of simultaneity, which represents the present. The change in position of any elemental particle can give way to the informational concept of time.

Does eternity exist? This is a concept that comes from asking ourselves whether something could exist indefinitely. If time is an informational issue, it does not make sense to take it as a reference to matter. The point is that we create the past with memory, we create the future by thinking about what will happen; both are informational concepts. Then we extend the notion of time and talk about eternity, only the present exists. We need to know more about matter; we are using a very large scale by taking the atom as reference. Now, if we take as reference the elemental particles, the scale increases. We are not aware of our structure and hence time, as an informational reference, serves to create another informational notion, eternity. Matter does not age, matter does not deteriorate, and matter or beings are transformed when elemental particles change position. This enables the creation of information, level after level. Each level of the structure of life has its reference to motion and information, but ultimately, particles changing position is what exists and with that, humans have created the notion of time.

The idea of the future takes very interesting personal connotations; can you tell me when the future starts for you? The future is something that is a second, minute, hour, day, week, month, year, five years, or perhaps a millionth of a second ahead. Remember that a second is a fraction of the rotational movement of the Earth. It can be measured by referring to the sun or stars. We think of time in many ways, all referenced to positions which relate to events. Together, positions form a transition. One second is equal to 1/86,400 of a revolution of the Earth. One second = 1/31,556'925, 975 of the length of the 1900 tropical year. (Alonso & Finn, 1992)

In summary, the model of time requires a reference to something. It's like any object that we create in our mind. To create this concept, we use memory; the least possible retention of information can help create the notion of time.

10. The Information Trap.

There are many informational traps. Two information traps which we can fall into are egocentrism and illusionism. Egocentrism is when we think our particular truth is the only truth that exists, our philosophy is the only one that counts, other people are heretics and their truth is wrong. Illusionism is when we think matter does not exist, when we think we are ethereal or we can live without the support of matter; that we can live only in the world of information.

The first case, egocentrism, embodies the authoritarian. Bacteria represent such an idea at a micro level; they are independent. At the human level, the information trap starts in childhood, when our fathers, uncles, or others begin to inculcate ideas we do not accept, but we are forced to accept. I'm talking about abuse, not a healthy relationship. In healthy relationships, the elder respects us and we are required to respect them, we are taught to be responsible, they insist we have discipline with our goals, they show us the benefits of teamwork and they nourish us with love and support. In abusive relationships there is no respect, children become workers at the parents' will, and sometimes become, in extremely abusive cases, sexual partners with or without kidnapping, but surely as slaves, and all this is in the twenty-first century. When these abused children grow up, they have a distorted view of relationships, respect, and what love is. We have seen that many serial murderers have had childhoods of fear, abused by one or both parents. Children are the future; we must educate them to escape the information trap and have an intelligent culture. This not only corresponds to the state, but also to us citizens. In summary, nobody can handle everything by itself and the minimum you need to exist is matter.

The second case, illusionism, embodies ideas of external control. There are people who believe that matter does not exist, that everything we see is imagination. In other words, they are imaginary beings, living imaginary lives, and their existence is imaginary. A model like: what I think is what counts. Others, accepting matter, say that the particles of matter are so small that they are not worth considering as real things, and jump into ideas where only the imagination has a place, thinking we are the creation of the imagination of other beings, or of information itself.

Let us review concept of illusionism. A person may think all that exists is imagination, and that everything happens according to our imagination. "I can imagine going to the street and being thrown into a car where I die; so I died and that is it; imagination is everything. The power of imagination is unlimited." After thinking for a moment, someone can ask: "If everything is imagination, can you sit where I am seeing you and imagine that you are thrown at a car and are hit by that car, while I see you sitting in front of me?" A little difficult, but if you do it, then you will convince me about what you say. You can invalidate the illusionism idea otherwise. You think you do not need to eat or have food of any kind and because you are the product of imagination, you can keep the same health and weight, simply because you think so. If you start practicing with your imagination in this way and stop eating, for how many days can you imagine that?

A test, a simple material but effective way to understand the information trap, is to put your finger in your mouth and bite hard until it hurts. If it does not hurt, you can do it until at least your finger bleeds. If blood comes out and does not hurt, you could still imagine that there is nothing wrong, despite the bleeding at your finger tip. Note: there are people with problems in the afferent nervous system, who do not discriminate pain, and for this reason, have problems detecting such biting or other types of incident; they perceive but cannot discriminate against those events.

I do not know if I helped you understand the information trap. The truth is that the process of information ability is not easy; despite my good intentions, there is no guarantee that you, if you are at some degree falling into the information trap, will leave it. Extreme egocentrism, or thinking that we're just imagination or energy, is an informational issue. The information trap is spread over many areas of life. We must understand that synergy is a lucid way, not an easy one, to achieve a better world and that matter is as important as information for well-being.

This concludes what concerns the concept of information. This approach, where information is something unique and individual, produced by an internal process from the being we call alive, is new (Arango J. D., 2010). It is still a long way for us to begin to manage it properly. The most important thing to remember is that this point of view should help us be more alert to the sophists, valuate our own thoughts as the source of our reference, and balance matter and information in our lives, essential for a healthy philosophy of life.

Third Part:

Life, Matter With Informational Ability.

"To live means to transform the dead and useless things into living and useful things." Giovanni Papini.

From the classical point of view "life is a property of organic beings by which they grow, reproduce and respond to stimuli" (Larousse, 2007). From a philosophical standpoint, "life is force or substantial internal activity, with which the being who possesses it works" (Lengua, 2009). For this book, life is a whole, system-process, synthesis from a specific material structure-disequilibrium, with which informational ability emerges creating an organism. That is, a material structure-disequilibrium that acquires the ability to discriminate the actions received from an environment and decides to act or not to act. The informational ability emerges (from matter) in a very basic way, imperceptible at our level. Then it is recreated again and again at different levels of living beings, reaching multicellular beings or groups of them. Informational ability helps direct the motion of matter, which in turn, helps support the process of life. The living process requires both matter as well as information. From the structural point of view, life requires controlling disequilibrium between the parts and the whole.

To understand a little more the process of life, let us look at the process of fire. Fire has been referenced as life for many cultures throughout history. Fire is characterized under two elements or components, a fuel and an oxidizer. While, next to each other or mixed, fuel and oxidizer will do nothing until there is an external condition, a catalyst, to start the fire. As an example, take a candle; the candle will be there until another flame turns it on. At that moment, the process of fire begins for said candle. The candle begins lighting. Paraffin is the fuel; oxygen from the air is the oxidizer; the catalyst is the same fire, which increased the temperature of the paraffin that was at the candle wick and started the process of fire in the candle and bred another fire. In this case, fire creates fire. Another example is how fire starts in the forests, there are certain conditions given from dry air, dry branches and dry leaves, and when lightning strikes, the process of fire starts. Again, the branches and leaves are the fuel, and oxygen is the oxidizer. The fire is an effect of disequilibrium. The fire represents changes to the structure of matter, and with this state change, energy is released as a disruptive action to other structures.

The energy released has several features, so far, not clearly known. Light has different properties. Some are explained with the concept of waves and others explained with the concept of particles. In other words, we do not know if light is caused by particles or caused by waves, although we do see the results. In terms of our model, we know the system in its initial form, the candle or the forest, we see the transition, i.e. the process, and then identify the system in its final form, gases for the candle or ashes for the forest. The analysis of initial and final systems i.e. structures of matter involved before and after this process, are known at the atomic level, but the very process, what occurs in the fire at the atomic level, is unknown. The same applies to the process of life. The chemical composition of the human body is very clear; we see the actions between us and the environment, the change on it, but most of our internal processes are a mystery, particularly the informational ones. Why or how is the phenomenon of life possible? Like fire, life starts from life, in a process that we understand requires information. Also, this process of life has a beginning, quite possibly more complex than that of fire, but the beginning is natural, not supernatural. It starts due to the many actions of matter, given the right conditions for life; it is a beginning we do not understand, but the process is there. The lack of knowledge about the process of life is such that we talk about the origin of life dating back billions of years, when likely today, with the number of living creatures that exist, elementary forms of life should be emerging.

Then, life is a material process as mysterious as fire, which allows a group of particles in a highly improbable dynamic structure (Bertalanffy, 1976) to perform several processes, the most essential being the creation and management of information. Think about the elements of living beings, matter and information. The material disequilibrium does not mean anything to the inert matter; however, it is essential for the creation of meaning, as mentioned in the concept of perfection, equilibrium, and other basic informational concepts. At its most basic level, the disequilibrium of the elemental particles is only a change of position –that is a fact; but how does that change begin to be discriminated in the structure of living matter? These are important technical details, yet to be resolved.

For you or me to perceive and discriminate a fact, it should involve an adequate number of particles, i.e. the number of particles that have been disequilibrated must be relevant or there will not be enough energy to reach the threshold required for you or me to perceive. In other words, if the fact is not significant to our structural level, let us call it our normal level, we will not perceive it and therefore we will not discriminate it. There will be no informational process generated within us to reach our level of perception. The living being maintains reference to the perceptions, i.e. memory, with changes to its structure; this is the ability to accrue information. But at the same time, at lower levels within us, other living beings use matter to accrue information, such as cells in the genetic code. An example is the change in position of an atom; it is imperceptible by us. But the movement of an atom makes a molecule different from another, and if it carries a message within our body, the message is different with the single change of that atom. However, the change of atom is not perceived directly, only the change of meaning for the organism is.

This tells us that life is a process with matter-information-matter actions or matter-matter actions in order to support the informational structure. This process is created by actions across levels and/or at same level of the structure. At the "initial" level, a group of prokaryotic cells communicate among themselves, and then according to the meaning of this message, the eukaryotic cells will broadcast it, and so on, increasing informational sophistication until the brain interprets it. In other words, the increase of informational ability is given by a process that refines the fact perceived through the body into "better" information until a decision is made and the process goes backwards, acting on the fact perceived. Still, others see it as the ouroboros, a self-absorbed process, see Figure 7. Dragon Against Dragon, where the boundaries between matter and information, or between living beings and other beings are blur, making it difficult to define what a living being is.

Figure 7. Dragon Against Dragon.

Let's summarize up to here: particles, creating limits. Particles are matter, part of them build living beings which create informational limits.

Matter exists and the informational ability of organisms defines limits. Living beings, which are part of the group of particles that exists, have informational ability. Life comes from below, emerging from material processes, from its structure; the basis for life is matter disequilibrium. Information is created by the material structural-disequilibrium of living beings. Ultimately, informational limits are being "created" by every living being discrimination. Without living beings attributing meaning there would only be particles in motion, nobody would define boundaries that define objects. No living being may create objects from the outside for another living being; it would be like seeing for the blind or hearing for the deaf. Thinking is an individual living-being activity. Look at it this way: there are particles. A group of them have the ability to assign limits and in the allocation of these limits information is created; the ability to assign limits and create objects is the informational ability. If there was no way to assign limits, there would be no way to create information, and there would be no reference by which to define the objects. Limits define objects, and structures are created with them, i.e. systems are created. The systems are informational creations, models of reality. By comparing the object saved with the object at hand, if differences are identified, those differences are giving way to the idea of the processes. We are saying, "This system changed, what I have memorized is not the same as what I am seeing. There was change, I am discriminating a process. "

Let us take a human being as a reference to see different structures.

Under the point of view of system:

1. One human being is a part of the humankind system (meta-system).
2. One human being is a system.
3. One human being is made out of organs (sub-systems).

Under the point of view of the process a human being performs:

1. Material processes.
2. Informational processes.
3. Combined processes called life, made out of material processes and informational processes.

We can also look at the structures of parts of the human being, from a "classic" point of view, as parts grouping together on three levels:

1. Prokaryotic cells.
2. Eukaryotic cells.

3. Human being.

The stellar structure of a human being, by decomposing from the whole, humanity:

1. Humanity.
2. Races of human beings.
3. Human being.

The structure of the living being of a human being as a system, parts interacting:

1. Cell systems (which form an organ).
2. Organ systems.
3. Human being.

Remember, I use my points of view, but if you use yours in your informational process, you will find that there are also structures; objects are contained within objects or objects are beside objects.

Looking at the leadership point of view, we can say that we the humans are structured as follows:

1. Leaders.
2. Followers.
3. Spectators.

Here we are using a concept of movement -behavior- which is the characteristic of a part from a system according to the dynamic it performs, cybernetics. In this case the driving force is information. If one human being is followed by others, he is a leader. The one that follows is a follower, and the one that does not follow, nor is followed, we call spectator. This is using logical rules, to be or not to be. Now, you can be a spectator socially, but may be the best leader for yourself. You use your good judgment and don't have to follow others. You are independent and judge each case at a time. From different points of view, you can be a leader, follower, or spectator. When you are invited to join a group and decide to be part of it, you are a follower. Also, you can form your own group and in this case, from the point of view of being the one who created the group, you are a leader. There are views that do not interest you, perhaps politics. Someone invites you to join a political group and you say that you do not like politics. You do not understand that it is the most important team, the team that defines the rules, so you do not participate, and you become a spectator.

Politics is essential for humanity because is the way to create alignment to an objective and the media to control both. A humankind objective should come from the humans' collective will, humanity survival. Alignment should let all live and support the collective defining ways to maintain individuality. The alignment need to start from within the families creating synergy. The control should be based in the group results, with proper equilibrium between needs and desires. Let now review these political concepts.

Life Objective:

We have talked about structures, change in the structures, viewpoints from the top, bottom, etcetera, but we have not talked about direction and sense, an objective. An objective is a decision about the direction and the sense where a living being should move. An objective is an informational issue defined only by each living being.

The objective of the living being structure is to live. The complex structure of the living being seeks to maintain a dynamic equilibrium among its parts and for that, information is used to discriminate what is useful and what is not useful. By useful, I mean the material elements required for life, which provide the energy to continue the living process; we are referring to food. The food, that for us includes water, still requires oxygen to support our process of life. In other words, if we do not have information, there is no way to discriminate if something material is useful or not in maintaining the living process.

But the living being is in an environment where there are changes which are not in its hands, like spring, summer, autumn, winter. Some represent wealth, and others represent scarcity. The process of living must continue in these cycles, and what we find useful in one cycle may seem useless in another. One example is that your body stores fat. It is not necessary at abundant times, but is invaluable in scarce times. The body has some trouble storing the fat, mostly because it is difficult to transport it everywhere, but at some point it uses the food as reserve in the cycle of scarcity. At this point the body must make a decision on how to manage resources.

We said that the human being is a living structure that has emergent properties at each level. It is difficult to travel through the levels of the structure and also across the structure at each level. Every living being is established as a black box (Ashby, 1957). Every living being performs actions that have to do with maintaining the structure; we usually refer to this activity as the operation, a process to maintain the state of the structure. The living being moves according to its possibilities, in search of food. There is a function and an objective. Other structures of matter function but do not have objectives.

Then, to live is the main goal of life. This objective is framed by disequilibrium, both external and internal, and in each case, both material and informational. By issue of animal structure, the mind is the informational sub-system out of the animal system. In human beings, the development of the mind has led to the creation of the consciousness of our existence. The mind is the only item that is in our hands to control. Some refer to that control as attitude, the direction and meaning that we give to our life. The proper informational equilibrium is transmitted over the material processes, creating a cycle where the proper use of information allows us to attain material results, which in turn, support our information or attitude, self-esteem. Good or bad meaning comes when comparing objectives to results.

Life Control:

Living beings use information to maintain the dynamic equilibrium in the processes of life. There, in these processes, we found the classic concept of control, to bring a process into a specific direction established by the objective. The process of fire is given only by the material characteristics; there is no information directing the process. If there is no fuel or oxidizer, there will be no combustion regardless of the temperature or the lighting. In the living, the process also requires matter-energy within acceptable limits; these limits are not determined by the living; these material limits are set by matter. In other words, too much force over a living matter can destroy the material structure that makes that matter a living being. In this material event the living being has no control whatsoever, is in the hands of nature. The informational ability allows us to find those material limits, where the process of life can be destroyed and brings the existential concept of fear.

We living beings define objectives with our informational ability, which allows us to extend the domain of the process of life and look beyond our immediate environment at impossible distances for other material processes. But more importantly, using the same informational ability used in the detection of energy sources, we set common goals for other living beings, and structure groups of living beings with common goals. Then we use the informational ability to monitor and "control" the new larger whole, the new living being formed by other living beings.

So we can say that the control of living beings goes beyond the functioning that matter has due to its basic characteristics, mass and charge. The control of living beings is based on the material properties of its structure, informational ability. Example: living beings create replications of their structure in a unique repetitive dynamic. Thus, living beings use control over functions to grow, reproduce, self-regulate (homeostasis), feed (select other structures of matter as energy), and flee (respond to changes in their environment). Living being uses control to maintain a material disequilibrium very unlikely for other structures of matter. With this, these structures of living being are creating a process that is not seen in any other structure, a process in disequilibrium but under control, dynamical equilibrium. The control in the living structure is given by the whole, matter-information; it leaves the simple material action-reaction to pursuit an objective.

The control in the living being also extends to its exteraction with the environment. A living being that faces a disruptive action may obtain multiple reactions using information, controlling the space between action and reaction, such as when a pilot tries to avoid meteorological conditions that could destroy the plane. An obvious example will help. We have a cat and a diamond and we take a drop test. The cat with its skills and control, informational ability, usually manages to fall with its feet on the ground and absorb the fall with its legs, regardless of how you drop it. The diamond will fall according to the position we have it in before releasing it. It will make no effort to alter the action, the diamond do not discriminate or attribute meaning. There will be no reaction to the fall, and it will just crash according to its characteristics, and those in the environment; there is no informational ability. So control is necessary to maintain life processes, informational and material, and preserve the dynamical equilibrium that sustains the direction of life.

Life Alignment:
Comprehensively looking at the factors of objective and control, they are informational issues required to sustain life. We could say that our main goal in life is to live and the control should seek to keep all actions on this goal. Then, the control should help make

sure that any point of view used, material or informational, is aimed at preserving life. Let's call this concept or idea of control, life alignment.

Let's return to the meta-system, system, sub-system, concept as seen in the structures of matter. Human beings are systems from the meta-system humankind. Eukaryotic cells take the place of sub-systems from the human being system. They perform independent and interdependent actions without losing the individual character; adding the informational concept of aligning life, humans need to have a common goal. This common goal will add to the material support given by a rich economy; then humankind requires two viewpoints, informational and material. Since we need balance between parts and the whole, the objective of living, as humankind, needs to incorporate matter, information, wholes and parts. The meta-system, humankind in this case, should incorporate each of the elements, material support, a common goal, global order and personal freedom, in order to reach the goal of having a better life for humanity.

It is important to note that we have a tendency to see the world in one way at the time: concretely or abstractly, optimistic or pessimistic, etc. In a mechanical way, perceiving structures or describing the movement. But the ultimate cause for matter to "exist" is that we exist and give it meaning. We give meaning within the natural limits of matter, and when we cannot give a meaning, let us say coherent meaning, we talk about something beyond matter, something supernatural. Informational ability is not something supernatural. It arises from the perpetual action of matter that has mass and charge, is grouped in so many ways, and acquires emergent properties, leaving behind the characteristics of its parts to create a new whole. One of the most marvelous living beings to date is the human being. We have the ability to give meaning to the meaning of our actions, thinking over thinking. One issue is to do a job, another to give meaning to it. Placing bricks can be an arduous and meaningless task, or it may be the realization of a dream if we understand the objective of that action, to create a shelter (house, castle, or cathedral).

We are looking for the truth where it does not exist. We are seeking truth beyond the general truth, which is given by matter. We must understand that the nature we see is due to us, living being, because we are able to judge at our criteria. Those criteria may be seeking justice or looking for legality. The effort should be directed toward preserving the humankind, justice; not preserving the direction defined in the written word, legality. Both are paradigms created. Both are valid, specific points of view, but in working with them, they create different cultures. The first culture, based on justice, wishes to preserve humankind, giving more to those who contribute more and covering basic needs for those who contribute less. The second culture, based on legality, wishes to preserve the status quo, regardless of the contribution of the individuals.

For those who manage the notion of vectors and for those who want to learn, look at the model under the concept of vectors. A vector is an object. The idea of the vector has given way to a new understanding of physics and facilitates communication among scientists. We mentioned that the Euclidean Point is an object too. The vector defines the direction and sense in which the point in question interacts in the system. This interaction does not mean movement. A vector points to where the action of the mass is directed, does not require movement. When it moves, we talk about the inertia; when it does not move, we talk about the force. All this is abstract, a model. Parallel to this abstraction of vectors and the proper proportions, is what happens with our particular truth and the creation of information. We judge a fact from a point of view and we get a result. We judge from another point of view and we have another result. Now, when the vectors come together, or we judge an event from several points of view to make a judgment, we created an interpretation of a fact. If we use linear algebra, this would be such that every point of view creates an axis and the optimization function defines the overall objective; the result of the process is the interpretation. We estimate in a general direction and the sense can be maximized or minimized. Similarly, the general direction that will take the mass in question can be interpreted using vector calculus.

This comparison creates another step in the model for nature, a more informational point of view due to the abstractions, but more practical than the one presented before because of the linear algebra usage that allow analysis in a set of rules under a goal. This requires the application of the first model, system-process, and the rules used in linear algebra. To go ahead, i.e. to further abstract, including additional logic rules separates us from the probabilistic reality of particles and leads us closer to abstraction, to create informational limits that do not exist in reality. By making these logical processes, we move away from understanding that a single elemental particle in the right place, at the right time, can change the structure of a set of particles, what is called The Butterfly Effect.

Paradigms Of Matter, Information, And Structure.

A paradigm is equivalent to an informational point of view. In the system, all parts that constitute it are parts of the system. We cannot speak of a system and emphasize one of its functions, saying that a function is the most important. You are not you without your heart; you are not you without your brain; you are not you without one leg; in these cases you are another being. What we think and what we do does matter to us. Paradigms are viewpoints that do not exist in matter; they are in our particular truth. We use paradigms as conceptual filters when thinking; they become a way of processing ideas.

What we want to highlight in this chapter is that your objective and control over it become valid for you, because you rely on concepts that you accept, implicitly or explicitly. At the time you think, all of them shape your particular truth. Paradigms can be based on concepts of matter, information, structure, or any particular object-model you want. You decide that.

Observe how paradigms work with three general concepts, material, informational, and structural.

Material Paradigm; Practical-Idealist

This paradigm focuses on seeing through matter, the result of acting. Here are two extremes, one practical, and the other idealistic. Seeing through matter, the practical is rational, what is needed, what works, and what gets results. Similarly, seeing through matter, the idealist is emotional, sees beyond what is necessary, sees the beautiful, and doesn't settle for what works; it must be perfect.

Look at the example of two farmers. Each one has a parcel of one block.

The practical, let's say, is aware of his limitations and lives from his land. He plants a fraction of his land, and has a couple of cows and some chickens. He lives from it, has some savings, sells part of what he sows. If he would like to grow, he would do it little by little, let us say in a conservative way. He does not risk what he does not have, does not take loans, etc.

The idealist, the one of which we speak, is not aware of his limitations. He is looking to plant one hundred blocks. A couple of cows are too little and taking care of a few hens for some eggs is not worth it. He is making plans to become rich from the one hundred blocks that will be planted. For that he is not doing, he is thinking about planting these one hundred blocks. He must convince others that he can make it. He will get results if he can communicate his ideas and have someone to sow for him.

Informational Paradigm, Pessimism-Optimism.

This paradigm focuses on looking through information when thinking. In this case, we discuss the abstraction that we make about the future, the plans and its risks. We have two ends, one pessimistic, one optimistic. With the same information from the economy, the pessimist makes abstractions and thinks that everything will go wrong, does not want to take risks, thinks that everything needs to be perfect before starting; he can keep making plans. Similarly, the optimist, in making his abstractions with the same information from the economy, thinks that everything will be okay, wants to take any risk, and doesn't analyze so much detail before beginning because he is confident that everything will be fine.

An example, two children receive an inheritance and have little training in their parents business.

To the pessimist who receives the inheritance, there is no satisfactory way to move forward. His conclusion might be that he must sell everything received. He will not be able to get anything out; he does not know the business; he cannot learn; he has no luck; he will sell to the first buyer. He does not want the whole business to deteriorate, become unattractive to buyers and then be unable to sell it.

The optimist, on receiving the inheritance, will think that the business is easy. If his father could manage the business, he will do so. There cannot be problems too large to be resolved, the business has been prosperous; he will make the business work as never before. As an extreme optimist, he will trust proposals to expand all areas of business, as if he had unlimited resources. Everything must go well; he will create the world's largest business.

Structural Paradigm; Analyst-Systemist.

This paradigm focuses on looking through the structure upward or downward hierarchy, from a reference level. In this case, we have two ends, one systemist, and another analyst. The systemist finds how to synthesize the elements; he sees to create wholes or systems reaching up to the universe. The analyst performs the reverse process; he breaks down or separates wholes up to the natural undivided, the atome[13].

An example: Two people are looking at a man and a woman walking side by side on the beach.

The systemist will maintain that they are a couple. There is interdependence in thinking, one limits or expands the thought of another; without one or the other, there is no system; they are interdependent. There is no human procreation without the combination of 46 chromosomes; this element of two parts is a couple.

The analyst will argue that two people, free thinkers, are a man and a woman, they are two individuals, they are neighbors. There is no way that they are a single element; an element from two does not exist. They are two individuals.

Note on the dynamics of a couple: In my house, my wife knows more than I do. I'm not sure if this happens at your home, but in general, I have noticed that women know more than men. Look at it from the viewpoint that the man proposes and the woman decides, then, having the right of the final move, there defining that in that team they have the last word, and so the confirmation of knowledge.

---

[13] Remember that the semantics of word atom is lost. Today an atom not only is an atom according to its semantics; the atom is now a name for a set of atoms or fundamental particles.

The analyst finds how the parts from the whole work, create components, in the process where he reaches the atom. With as much informational ability and the analytic approach within the structural paradigm, the analyst will find key elements in the operation of the whole at hand.

Let's look at an example of why it is important that the whole structure works well, that we all do our work in a professional manner.

Reading a magazine that sells machine tools, I found an article that shows how a part of the system, no matter how small, can ruin a huge effort. One day in a decisive battle, a small event took place; a nail came off. The nail was holding a horseshoe, which came off. The horse that had the horseshoe lost control of itself, did not obey the instructions of its rider. Its rider was distracted by the actions of his horse. The distraction of the rider served his opponent to strike a mortal hit for him. The rider was the king. This led to confusion among the soldiers, who began to fear for the outcome of the battle and were disbanded. The decisive battle in the war was lost. The war was lost. Make an exercise of this passage. Think about what could have happened wrong. Think about what you would do for that not to happen.

Then, paradigms do not change the facts, the structure of matter. They are informational issues, elements of particular truth, which are usually applied without consistency. That is, sometimes we look at what we do, efficiency, etc. and sometimes at why we do it. We carry out some projects where we fall short; we were practical, tactical, etc. We do not extend the project with a bit of idealism, do not think strategically. Sometimes we think that all our opportunities are finished in an attack of pessimism, or the contrary, we think naively that no factor can destroy or damage us in blind optimism.

Look at the general paradigm presented in this book: information is the result of the informational ability of living beings. There is no information out of the living, which is matter with informational ability. A book is considered information; but it is not so until the reader is illustrated. Failure to reach the goal of enlightening the reader has many possible causes. Among them, the writer does not write at the level of the reader or the reader is not at the level of the writer. With another paradigm in mind, this book seeks to develop faith and courage; faith that the world can be a better place and courage to fight for a better world. Justice is an adequate paradigm for a better world, but it requires the proper management of information with the faith that justice can be reached and the courage to implement it. We make mistakes and it takes more courage to accept them, justice, than to let them pass, peace.

Thinking, Movement Of Information.

Paradigms and other models are our particular truth. Combined, they form our information system. Moving from one paradigm to another, reviewing models, becomes an informational ability process, thinking. In this case, our mind is making the process. No matter where we are in the brain, we are within ourselves; we are in our particular truth. Imagine you're visiting a house. You go from one room to another but remain in the same house. This is an abstraction to the way our mind can function, or the ability of consciousness in each of us. We go from one part of the brain to another or from one paradigm to another; we are always in the same house, the brain. Then, thinking is moving information or moving through our information.

To think is to combine or recombine direct discrimination or information accrued in the brain. By thinking, we are creating new ideas. There is no absolute reason for this; it may be curiosity, the desire to roam our ideas or feeling a need to satisfy. Thinking is a permanent brain action which requires conciliation or repair periods.

These periods are usually within the sleep time. Then, by thinking we are moving ideas from one place to another in our brain. The brain is the system –the mind is the process.

We speak of equilibrium as a basic information concept. We have also discussed that matter and information go together in the living being, without one of them life won't exist; that's an equilibrium to be controlled. But these ingredients are not next to each other, they are integrated into another structure, which has been mentioned as a whole-part and has also equilibrium to be controlled. Both matter-information and whole-part equilibrium have a need for control that is imposed from within. Again, this balance is characterized by the dragon that is eating its own tail; see Figure 7. Dragon Against Dragon, being represented in many cultures (Livas, 2009). Life is controlled from inside.

Material changes, defined by other structures outside the living, create limits which coerce life. They are an environmental reality, the system's environment, leading the living being to interpret it. As we have said, the interpretation of the matter around us, according to our informational ability, creates objects that represent nature. In this case we mean that the paradigms are an informational reality, creating a complement to that material reality. One example for humans, where there are several interpretations of that informational reality, is shown in the book *The Four Giants of the Soul* (López, 1965) Fear, Anger, Love, Duty, are informational elements which do not exist in the general truth; they only exist in the particular truth. These interpretations that were created when thinking are merged into the information of the living being and become the guide to its life. The informational elements are allowed to control the specific objectives of every living being, which are born from its particular truth, versus the collective objectives, which arise from the conventional truth. Because of fear or duty we stop doing something, or vice versa, we perform some actions because of fear or duty; the decision is personal. An example is that of an informational disease, depression, where an individual is trapped between doing and not doing. If the individual does, let us say defends him from an attack, wounding his attacker, he feels bad, and if he does not defend his position he will feel bad too. Another example is when a person is told that what he does is right and the person does not believe it, having no source of pride; or the person is told that what he does is wrong and the person believes it so and then has no source of pride.

Scientists know about the electromagnetic capabilities of some organs in the body, in particular the heart and the brain, a material issue with informational consequences. What are they doing there? The heart seems to have the strongest magnetic field in the body. The brain seems to follow in this context. This aspect of magnetism has been mentioned, because until now there are no detailed studies of these magnetic fields. We use them in electrocardiograms to see heart performance but do not understand if they can be used by the body to communicate among organs. What those magnetic fields contain as information and their possible uses in the integration of our informational ability needs to be understood. We are not talking about what scientists think but about how these fields are used among the cells that make up our body, or the colonies that form the various organs of the body. It is curious that these fields are used only for broadcasting. Is there a code in them? Is it maybe the way neurons use to communicate and synthesize objects? Electromagnetic links are commonly used in communications and our primitive devices handle several options, but we have no electromagnetic-cell code, or the intensity is very low. We're probably like the dog in the section that describes the fourth step in the information process, comparison; we do not know which the cause of pain is: the car, the person, or the cane.

Let us try to integrate the elements of life, information-matter and whole-part, in a brain model developed by Dr. Katherine Benziger. As noted, the brain can be seen as a home. Say that house has four rooms. These rooms are contiguous; see Figure 8. The four parts of the brain according to Dr. Benziger are four sub-brains, and each is like a room having information processing beings, which are processing what is going on in the streets. The front rooms only handle informational paradigms; the rear rooms only handle material paradigms. The rooms on the right work to integrate objects, and rooms on the left see the parts of objects and its parts. The Benziger model is a very rough approximation of a described reality, but we must start somewhere. As in the classical model, which is widely accepted by psychologists, there is one rational side and there is one emotional side; each person has a preference, which we presented in the basic informational concept, points of view. Benziger says we can also have preferred ways of thinking, not two, but four options. The most significant difference from the classical model is that the preferred mode has a physiological explanation, or is based on material conditions. Before discussing the preferred mode, the law of dominance, look at the model.

This model, created by Katherine Benziger, is based on many years of study and practice in the area, or let us say life point of view, of psychology. The studies include analysis of brain scans to support her work. In the model, the whole is the person; one individual part is its nervous system. In turn, a component of the nervous system is the cerebral cortex, or neocortex (according to Maclean). The neocortex is responsible for the functions of thought, the informational ability of human beings. Benziger denotes four modes of thinking. Without departing from the traditional model of two lobes, where each takes on rational o emotional functions, the model adds the concept of abstraction and concreteness for each of the lobes. The basal parts of the lobes, to the back of the individual, are the beings who process the data received from the senses, creating objects, masses, concrete images and first-order information. Then the front parts of the cerebral lobes, defined by the central fissure, process the already discriminated by the basal parts once again, creating new abstractions, second-order objects.

In her book, see Figure 8. The Four Parts Of The Brain According To Dr. Benziger, we draw two axes, one in a forward-backward direction (longitudinal axis) and another in the left-right direction (central axis), forming four parts of the brain like this:

1. Front-Left (Brain front left)
2. Front-Right
3. Basal-Left (Brain left rear)
4. Basal-Right.

The physiology of the brain, another name for the disequilibrium of brain matter, gives the possibility to one human being of having different ways of processing reality. A fact can receive four different interpretations; the informational process that each of these brain parts made, creates four different informations, i.e. discriminates a fact in four different ways. They create four realities that your mind finally assigns value to, in other words, your mind makes the final interpretation.

According to the axes mentioned above, each sector performs like this:

1. The abstract parts and key concepts, **Analytical**.
2. The abstract of the whole, symbolic, unifying image, **Synthetic**.
3. The parts perceived or tangible objects, **Structural**.
4. The connection or interaction of the perceived, **Relational**.

The preferred way comes from the dielectric capacity in the brain, see:

> "*According to research done by Dr. Richard Haier in at San Diego, we 'prefer' one mode because our brain is*

naturally more efficient in that mode. According to Haier, the 'electrical' resistance within and between neurons in our area of preference is so much lower that we only use 1/100th the oxygen or energy when we use it to think. In other words, when we use our natural lead, thinking feels easy and effortless. By contrast, when using our other areas, each of which uses 100 times energy – thinking is literally more difficult and exhausting.

What is important to note is, physiologically, dominance is natural and normal. As a matter of fact, dominance governs much of our physiology. And yet, as natural as dominance is, it is frequently not understood or not accepted as valid and its implications are often ignored. Moreover, this simple lack of understanding results in many people finding themselves confused, tired or overwhelmed by problems and block from the joyous, effective and healthy life they seek."

(Katherine Benziger, 2006)

Here again, matter is the basis of the informational process, in this case, as a preference for the use of our informational ability. We can say that the dielectric properties of the brain fluid define a comfort zone when we are thinking. Usually, people feel more satisfied when working in accordance with their preferred mode; it is easier to work in those areas where our brain moves more easily, or it's easier to think. Just as there are people with a preferred mode among the four, there are some people at the other end, who can use the four parts with the same ease. That is what we call full use of the brain. This should clearly show us that combining levels from each sector forms our reality, our true nature, making each of us unique because of the endless combinations. This can lead us to say that happiness depends on each of us, since everyone has their own expectations of success.

When operating on the law of dominance, Benziger declares two empirical rules which help us to live better: self-esteem and survival

1. To develop and nurture self-esteem, because as we work where we are productive we will gain independence and value ourselves more, and

2. To ensure survival; as we further understand what we do and further understand about other people, we may improve the interdependence with other people who think differently from us.

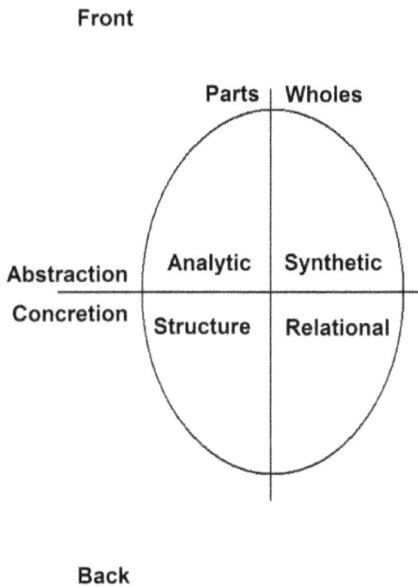

Front

Parts | Wholes

Abstraction | Analytic | Synthetic
Concretion | Structure | Relational

Back

Figure 8. The Four Parts Of The Brain According To Dr. Benziger.

Here we are going back to life's common elements, matter and information. The brain, food, survival, etc. –material elements to the system of living beings; and the mind (informational ability), meaning (information), self-esteem, etc.–informational elements to the process of living beings.

Chapter 7.

Uses For Information

"Enlightenment is man's leaving his self-caused immaturity. Immaturity is the incapacity to use one's intelligence without the guidance of another. Such immaturity is self-caused if it is not caused by lack of intelligence, but by lack of determination and courage to use one's intelligence without being guided by another."
Kant

The main use of information is to support informational ability; that is, making decisions to support the process of life. The living being, to meet its vital functions, requires accrued information, which enables the living being to maintain the condition that characterizes its improbable material state. Thinking is a process. As we saw, learning is the result of the application of thought. By learning, we move our particular truth and with that, the reference for thinking from one place to another. We are on an informational path, ontogeny. Comparing what we think with what we thought shows us the movement of information, which is the path we have traveled. Children do not understand the behavior of adults. Adults forget some elements of what they have learned as children. It is interesting to see the process of learning and unlearning through life, the change in the way of thinking, which confirms that thinking is moving information. Let's look at Figure 9. Thinking Under Four Points Of View In Life, showing material, spiritual, self-affirmed and instinctive trends, developed in the process of life. Figure 9 is taken from Pinillos (Pinillos, 1970) but is originally Moers'.

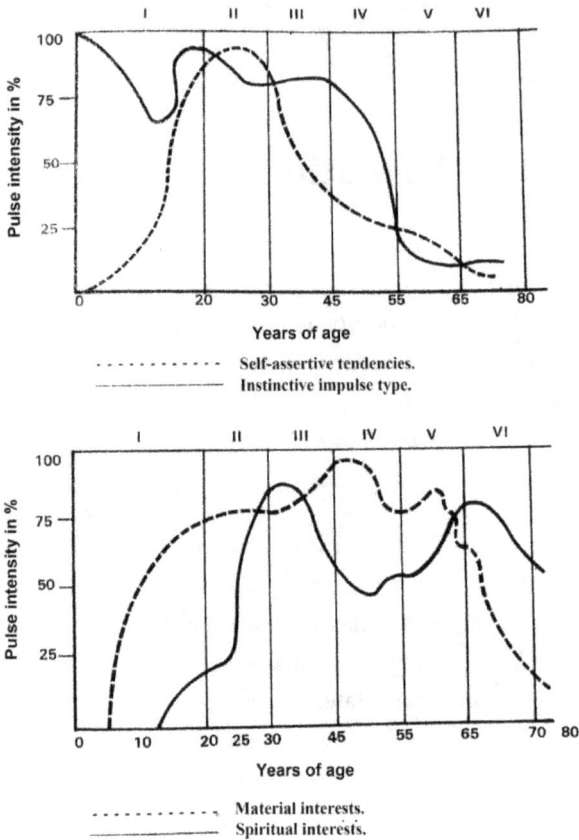

Figure 9. Thinking Under Four Points Of View In Life.

In living organisms, the informational ability with its accrual capacity can create what some call Anticipatory Effect (Rosen, 2003). Information accrual or memory allows the living being to use its experience through different ways. An example is seen in the genetic code, where accrual is used to pass on experiences to other generations. Decision making is at the center of the use of informational ability of living beings. When making decisions, we are judging the usefulness of a fact. We compared the odds of that fact with the goal of living and other elements that we have within our particular truth. To decide becomes a plan, because we are thinking with an action in mind what is best for us in a moment of reflection, or in a moment of urgency. In any case, this is a matter-information process that is improved with more informational ability, in practice-theory cycles that bring us learning.

Informatically, thinking-deciding-thinking becomes learning. By combining objects, we create models. A model may become an axiom. An axiom is still an object. The axioms create the basis for informational structures. The point and the vector are axioms for mathematical models. We are building informational structures where everything has meaning and is information; we do not build anything material in our environment. Without implementing, you can apply that thought and create more ideas, until you reach the idea of perfection. Perfection is an object created by analogy to other material objects, of course, it becomes the reference model. We reached it by making decisions, which we call logical, there is only information.

Materially, thinking-deciding-executing also becomes learning. The particles create the basis for the material structures. When we combine particles according to their properties, we build structures. Because the structure of the particles or the change in the structure ultimately defines what we perceive, we say that general truth is given by matter. In each cycle, thinking-deciding-executing, we perceive different material structures. When reviewing what has been executed versus what was decided, we can find that there is a difference between what we think and what we execute. When that difference is reduced, is what we commonly known as learning. Then we can say that to learn is to correct the planned and/or executed to be more consistent, improving the chances of getting the results that we want. That is the essence of total quality control to reduce differences, what we call mistakes; the fewer mistakes, the better the quality. By analogy, if our results do not meet what was planned, we are moving away from what we wanted, what we liked. We would be unlearning; the quality of our actions is getting worse.

Economy, as a model, seeks to decide on the use/allocation of resources with the aim of obtaining maximum benefit. The maximization of resources is a principle used by living beings in all their processes. Nature interacts under the properties of matter, inertia and charge. There is no way of creating or accruing information without the actions of matter; then, it is necessary for living beings to make decisions in the most economical way possible, taking us to the concept of minimum effort. Now, the most economical way possible is under the particular truth of every living being. It is a comprehensive decision which includes the two basic points of view, the material and the informational. Making economical decisions, i.e. maximizing the results with the resources at hand, is seen in most living enterprises.

Decision Making

In its most basic concept, a decision is information creation in order to support life. Let us recall that the principles are easy. Implementation is made difficult due to the number of elements involved and because each element has its own principles or properties. When performing the informational process, the judgment ends the process begun at discrimination. Every living being decides what is before its eyes, what is the sound at its ears, if the environment is hot or cold, if a food tastes good or bad, and decides its usefulness in terms of living. An example of this is in surrealist paintings. With a focus, we see a cup. With another focus, we see two faces; we are deciding on one of the two. In some paintings, we do not discriminate, our choice is: "I do not see anything." In this context, we are deciding on art, what the painting says. Art, beauty, are "reasons" for leaving. In the context of life, what we decide determines the outcome of the process of living; at our level, it is the combination of material and informational processes that allow us to live the life we want. That is, we are combining informational and material processes through all levels of our structure which begins in elemental particles. The decision processes emerge from within the structure each living being; starting in prokaryotic cells, having its internal decision processes that give way to the eukaryotic cell; the eukaryotic cells decision processes give way to the organs processes; organs processes give way to our body, where ultimately our mind decides on the interpretation of what concerns us.

When we make decisions, we make them as the whole we are, that emerging whole from the material structure, the human being that we have used as a reference. Informationally, that whole can arise from the operations that are performed, by what that group of parts does. Example: one whole that writes creates one writer and so on; actors, negotiators, parents, etc. are created. When we combine all these functions and do so simultaneously[14], we have a new whole, in this case a human. When the time comes to make decisions, we gather more or less elements to make the decision, depending on our informational ability. But in each case, it is a holistic decision. If we consider which part of the brain we are using, from where we are gathering the elements incorporated, how we are considering them, and so on is generally indistinguishable. It's a whole which is judging, independently of the parties involved; we said that whole is our mind. In the unconscious process, that whole should review the four parts of the brain. But we talk of the preferred mode, which has less dielectric strength, which leads the economic process and because of that, the mind prefers it. Then, we are taking the decision as a whole, but in this case, we do not use the other parts of the

---

[14] I mean the fact that everyone plays several roles at once. You do not stop being a father when you go to work; you may have a couple of pictures of your family nearby. Same with your parents; you will remain the child until there are no parents, etc. Father is an idea, stallion is another.

brain with the same intensity. Our decision is not comprehensive, but still, we take it as a whole. This leads to biased decisions, thinking only of ourselves, only of others, only of the media, or only of the aims, which results in undesired consequences for us and the groups we are part of.

Theoretically, the integral use of information would have all points of view, at all levels, at all distances. The issue is practically impossible in reality; so, to start a process of practical decisions, we define points of view. First, we have matter and information, which for human beings can be seen as needs, from the matter side, and desires, from the information side. Note that these two views may include those in the Maslow pyramid of needs. But in this case, there is no concept of a pyramid, there is a symbiotic concept, the ouroboros. The decisions are proper to the living being that makes them; it decides whether to satisfy a need or desire. It is its choice! Sure, some may think like the character of a fable, when at the moment of an assault he is asked, "Your money or your life," he responds: "My life, I need the money for a business." Decisions of living beings seek to satisfy their needs, but it is not necessarily the case for everyone, particularly for the human being. Secondly, from the point of view of the structure, we have said that the human being is a whole that has parts and at the same time is a part of the whole humankind, and when making decisions, we should consider the other parties that make up the humankind whole. In this case, we added to the parameters of need and desire at the decision making moment; how does the outcome contribute to my life and to my community? Finally, we include the terms of distance. Depending on the distance, measured in cycles, there are two points of view, tactical and strategic. The tactical order is short-term, and involves things like hunger, cold, tiredness etc. The strategic standpoint, involves things like procreation, preserving a territory, protecting a drinking fountain or security etc. Let us look at how the integral decision becomes more complex with each element added.

Are there perfect decisions? As already mentioned since there are many points of view, the decisions that are perfect from one point of view, cease to be from another point of view. The possibility of making perfect decisions from all points of view is almost impossible in practice. Of course, we are not only free, but obligated to ourselves to seek the most appropriate decision, one that maintains integrity in the group for as long as possible. Group decisions are difficult when we have limited solutions, when we must decide in a few moments between extremes as to kill and to live, or not to kill and not to live. If you do not kill, you will be the victim of that opponent. But those are extreme situations, often from violent living beings, who live in a paradigm of scarcity, surrounded by soldiers with little or no criteria, soldiers obedient through pay or fear, mercenaries. They cannot understand beyond the extremes, do not understand that there are intermediate solutions which can create a better world.

How complex is decision making? Complexity is a concept relative to the level of informational ability. If we had perfect informational ability, there would be no complexity, there would be no chaos, there would be no problems because we would prepare in advance what action to take; everything would be fine. Finally, we would not have to think, just apply our accrued information, our knowledge. But I don't know anybody with perfect informational ability, who can track in advance all particles and their interactions. So, within our limits, decision making presents the need to discriminate based on the few elements at the time. The first discrimination should be made on personal goals and a second on collective goals, that is, we should have a personal goal and a collective goal. With this, we can define the necessary resources, both material and informational.

In a decision of trillions of living beings, there should be more complexity than in one of the billions of living beings. But if each of the billions of living creatures is more complex than each of the trillions of living beings, the complexity of decision making may be greater for the former. Not surprisingly, more informational ability would solve problems more easily. I mean, the trillions of cells in the human body form a very coherent whole. As humankind, the billions of human beings that we are, we work with the same concept of life as a pack of chimpanzees. The difference is technology, not the behavior as living beings. Then, from here the concept of informational ability can be divided between intelligence and consciousness, which we will discuss in the last chapter. We must improve our decision making, incorporating more professionalism, leaving aside the obsession for just the easy to do and finding the most benefit without sacrificing the future of other beings on the face of the Earth.

What is a blind spot or a mental blockage? When we arrive at a time of making decisions where there are too many items, or we surpass our informational ability's level of managing complexity –where we cannot prioritize– we say that we have reached a blind spot. Also, incorporating the emotional side, we can reach a blind spot when we are powerless, not by logical complexity but by an emotional issue. We cannot make a decision; or rather, we obstruct the decision because of the emotional aspect of it. This emotional aspect does not need to be a life or death threat. There may be "easier" elements such as not disappointing our parents or neighbors. Currently there is news in this regard. One nineteen-year-old committed suicide because, playing with his privacy, some colleagues filmed him with another man and exhibited it online in social networks. This boy came to a blind spot, could not go farther and confront his parents. It was easier for him to deal with the material action than to deal with the informational action. In other words, he preferred to break his integrity (committed suicide) than to confront his parents' criticism. The point is we need to try to see a little farther. We are worth as much as any other human being and we need to reflect on our true commitments and whom we have these commitments to –a nearby group, ourselves; or a distant group, family or friends, even a country as a whole.

How are confidence and clarity similar? In principle, it seems to not be a decision making issue, but it is like the previous point, obstruction. Clarity in decision making is having no blind spots, properly allocating priorities, being the first to preserve life, and in confidence you have no mental blockage, you can solve any situation without having to reach extremes like killing yourself. As we said, we have all our information, so in businesses it is necessary to keep in mind who we do business with. Exploring all the possibilities that are within our power in proportion to the type of decision to be made gives us clarity. And when it comes to business, understand which person or company you are negotiating with, it gives confidence. As we have said, there is no perfect information, but do you clearly know your business environment? What can you offer today that you did not offer yesterday? Do you want to keep your margins at the expense of your customer or supplier? In general, you should ask if your trading strategy is consistent with the mission and the vision that you have for your company in the market as well as of its the interior. That is, as expounded by Michael E. Porter, the strategy must understand the entire value chain.

What about risks? Every decision has risks. The risks come from the nature of nature, or should I say, from the nature of matter, including informational ability. That is, the constant interactions of matter and the communications between the living and the actions of those, which we call exteractions. When we make decisions we are predicting what will happen in a system, and excluding the possibility of other events. But we have no control, as we said; we do not have all the information that would be required to have perfect information. So the best way to "control" risks is by measuring them. Sometimes, we take risks like working with earth, creating a hole to cover another hole, when in fact we do not want either of the two holes –neither the one created nor the one that is being created! That is why it is important to define when we consider we are lost. That is, predefine when we must leave the path taken, the one we decided it would be and take a new path. In other words, we should be aware of the losses that we are willing to endure, the definition of defeat or how to define our defeat. Another point in measuring risk is to understand that we are developing plans, and all the battles are not won. We must stay on the course, but we need to understand what we are building.

As an example of risk measurement, in a negotiation class given by a Spanish negotiator who participated in the negotiation of the metro in Medellin, Colombia, he said, "When you are negotiating, you should know what you will give in return, even from the personal point of view, because negotiating boundaries are chosen by the negotiators." So, you may think your situation is hopeless because it has presented a negative risk, and at that point you can reach a mental blockage and make decisions that you were not willing to make, creating two holes and a pile; a complete failure compared to only one hole. Here, the point is to consider positive and negative risks that will allow you to obtain the foreseen system, or one of the expected wholes. In terms of investment, you lose the investment and not more than the initial investment. This is because many times, to salvage a reputation that does not need to be saved, i.e. trying to be seen as infallible, you will create more problems than you think would occur. Let us examine the collapse of General Motors, the car manufacturing company, which is seen as: executives departed from customers, successfully lobbied for Japanese cars to have more difficulty reaching the American market; then they bought Fiat, the Italian car maker, to remain the largest producers of cars in the world, and finally end up in bankruptcy. The simple solution, make good cars to the taste of consumers, was too much for the executives.

Decision making is analogous to the design process; they go hand in hand. They require prior information. Their objective is to seek the satisfaction of needs, with resources at hand or not at hand. For the two, you need to imagine future scenarios. But there is a difference: the decision-making process ends at the information side, while the design process takes the validation of what was designed and puts it into practice once the implementation has been decided. The design process still needs to deal with the material aspect.

Finally, it is important to mention that tools do not replace our role as decision makers. We are living beings who ultimately make the decisions, no matter how much technology is involved. In the case of the support of counselors, where the counselor provides a point of view, the user ends up deciding what is good or not. There are irresponsible people who insist on making decisions and then throwing the responsibility for the results to those who commented during the process. Books of quality control say that managers in Eastern companies take responsibility for their decisions, good or bad, while in the West, if the result is good, the manager takes responsibility for the decision, if unfavorable, the manager finds a scapegoat to justify the error. We always decide, even if we decide not to decide and give others the space to decide what suits us or not.

Communication

Communication is a process that allows the integration of living beings. As said, the use of information is to maintain the material integrity of the living being, the whole-parts, as stated in the decision making section, to live. But communication can be seen as the use of information to maintain the community, group of living beings, parts-whole. An example for whole-parts is: you are a whole, you look to preserve your life, you communicate within you to fight or fly, and you decide on it. For the parts-whole, an example is: you are an aggregate of parts, each cell has its own life and they (you) communicate to support themselves (you); and here, communication integrates or creates cooperation under a common goal. Without communication, as we said, there would be no living structures; there could not be teams of trillions of cells. It is hereby for the reader to decide which of the two uses of the information is more important.

We can say that we only have one use for information and that is to make decisions at the each living being level. The decisions of cells are reflected at the macro level by what we do and at the micro level by the actions performed by each cell. Let's look at an example. The idea of taking a liquid is created in your mind. It was a decision made by your cells. You decide to drink some water. When you move, your nerve cells are telling your muscle cells, "we are going for water," etc. If muscle cells do not contract, even if you want to drink water, you cannot do that.

Let us remember that the first forms of life date back a couple billion years ago. The integration of prokaryotic cells requires communication. Prokaryotic cells are nothing if the molecules that structure them are not there communicating.[15] We must clarify what the function of the ribosome is. How does the ribosome discriminate, accrue, compare and finally judge in the replication of prokaryotic cells? What is the code and the material means that the cell part use to communicate? These are all technical issues. The same principle applies to eukaryotic cells; the integration also takes communication. Having proper proportions, the integration of prokaryotic cells can be seen as a horse and its rider; they serve each

---

[15] This plain structure of the human being does not exist. In multicellular beings there are bacteria which are "friends" helping functions of the body such as the digestion. Also, with the help of fungi and other microscopic animals, we become a traveling zoo; we are ecosystems with living matter and inert matter in specific functions, as a support structure to the functions of life. An example is the enamel on teeth; it is inert. Others are the collagen in the skin or the body hair around the body, which are proteins.

info@matterinfolife.com

other and communicate through gestures or actions of dressage. In this context, the parts come first and the whole follows –living beings first, and then the communication in order to have superior structures, a greater living being.

Let us analyze the human being structure. We can see the human being system, the individual. First, we have a group of sub-systems, represented by eukaryotic cells that are groups of prokaryotic cells. Prokaryotic cells develop codes and communicate to create different types of Eukaryotic cells. Later, eukaryotic cells using communication ultimately create multicellular beings, among them the human being, which we see as one system. Now consider the meta-system of human beings, humankind. Humankind is in the process of improving communication among its parts. There have been various attempts to communicate through common objectives, with little success. But now, humans have developed material means such as electronics, and later other tools like the internet, which will let humankind improve communication, so that we will function as one living being, beyond what we are today. These living structures that represent humanity will continue to be complemented, because the process is already adequately directed.

The idea of control is widespread, but it makes no sense without the ideas of communication and consensus. When we analyze the idea of control, it is not easily clarified because deep down there is no control in the strict sense of being able to maintain a specific condition. When we say that a rider controls his horse, there is really a communication process where the horse, by the actions of dressage and/or care of his rider, accepts to follow his instructions. Communication in its best sense is a process that builds consensus among the living. Consensus requires that individuals integrate to the will of the team. A manager can communicate very well but will not reach consensus on the objectives of the company until all those involved understand that they are a team, which achieves results beyond what each one can achieve individually. In the team, each one performs a required function for the purpose of the team. The consensus is made under certain assumptions, most importantly, mutual support to achieve the team objective. The team is a system and its members or parts, sub-systems. The consensus is commitment to the team. It requires good communication, which is the information part, but ultimately, positive results are desired, which is the material part. Those who do not perform at the required level should inform the team of their challenges, and it is a team job to support other members that do not perform at the required level. The communication to reach consensus must be sincere, true, and understood. Only the communication, message understood, will enable consensus in the real world, in order to reach results.

Interaction is not communication, but communication has interactions. An example of interaction is a collision. One being collides with another and hence, there is an interaction, or action between beings. As a result of this interaction, depending on the physical characteristics of the beings and the state before the interaction, there will be a given result. Look at the interactions on the billiard table, particularly the regular billiard, three balls. Assuming standard conditions, the table is leveled, the balls are spherical, and the cue meets accepted specifications, etc. The billiard player hits the ball, and then there is an interaction between the cue and the ball. The ball is under the physical conditions defined by gravity, the cloth of the table, the roundness of the ball etc. In this result, inert beings have no chance to decide. The ball responds to the cue hit according to the stimulus-response physical nature.

In communication, there are also physical stimuli; the communication process has a space between stimulus and response. The response is defined as a result of understanding the message, the capabilities of living being and/or conditions of noise, not the various physical interactions that occur in the process, i.e. the code. If the message is understood, the sender and receiver have the "same" information, a concept was shared; after the communication, each living being knows what the message was. The billiard ball followed the physical stimulus and it will continue in the same physical way, without information, as result of the hit. We think, encode, perturbate the media, expect an answer that should paraphrase the message and if the message was understood, the communication step ends. If it was not understood, there is a new cycle to carry out the communication. I do not want to dwell on this. The point worth noting is that in the interaction, the process is purely physical, not transmitting any concept, because beings at both ends of the interaction are inert beings. In communication, information is transmitted using code, which will have more or less meaning for the recipient depending on its experiences (information); that is why when there is more knowledge shared, communication is easier.

Communication allows us to create conventional truth, "shared information." In cells, the level of communication is more basic, "simpler." For beings "close" to matter, communication is more concrete than in men, where communication becomes more difficult due to the higher level of abstraction which our ideas have reached. Technically, we define communication as a process where an information packet, which is the encoded message, is sent to one or more recipients, the recipient receives and returns confirmation of the information package received. In classical communication, the message does not matter; what matters is the material interaction, that the code is properly transmitted. Among the living, there are many material means chemical, magnetic, electrical, etc. but what is really important is the message, because it is a matter of life to preserve the organism. Upon closing the communication process in a cycle, the initial sender confirms that the message was understood when the receiver executes the expected action by the issuer. In this case, the communication is a vital process which seeks an outcome for individual and team reasons. The use of communication in living beings, once established, allows conventional truth creation and team work. In team work every part of the team is responsible for a vital function in the new organism, one of them being team coordination.

Let us emphasize that the code of the message can be received but not understood, having no meaning, not transmitting knowledge, and in that case, providing no real communication. In the classical communication theory, if the code arrives, it is said that communication took place. But the code only arrived. If you do not understand the message, do not understand the concept that is trying to be transmitted, then there is no real communication. When I tell you that a grain of sand of seventy kilograms can pierce your hand, when it is placed gently on your hand which is resting on a table, you can say you understand perfectly. Of course, it's possible. But from the fact that you say it is understood, to the fact that it is understood, is something else. You can repeat over and over again what I said. As a recording, you have recorded it in memory, but there is no connection with the material world. Communication among living beings should convey the meaning that we want; knowledge about the real world. Therefore, communication in the classical sense is strictly physical interaction, code transmission. Here we emphasize that it is giving meaning to what is perceived.

What is noise? In classic terms of communication, noise is pollution of the coded information due to characteristics of the nature of the media used in communication. In other words, noise is something undesirable which occurs in the communication process. In those terms, in pursuit of conventional truth to create a real human team, there is creation of many noise factors; this is what Covey calls the social mirror, what we are taught but does not suit our nature. Example: if we value peace, we find that when we are taught with violence to exercise violence, we find ourselves outraged. We do not accept violence as a valid factor for us. Or conversely, when we are violent because someone in our environment was violent, we find that peace is annoying, unconvincing. From the point of view of communication, noise is also given with misinformation. A leader shows how the world works for the benefit of all. A scammer misinforms us. He is trying to say that the world works for the benefit of all, but basically is just looking for personal benefits, not those of the community.

Is there a universal language? Possibly it is the brute force that generates fear, which is the language of tyrants; it is accepted by the group who does not understand diversity. The group wants to be equal to its friends, but wants the death of the "enemies." It is necessary to ask: Is nature our enemy? It is when it is not in "agreement" with us. Nature is the enemy we must master. We are fighting against global warming, one or two degrees Celsius in the last century. Why don't we talk about super volcanoes? Recent studies show that the super-volcano Krakatau in Sumatra exploded and changed the composition of the atmosphere, with an immense volume of particles. The particles reflected sunlight, cooling the Earth and changing the entire ecosystem, creating frost, destructing plants, and killing countless living beings. Researchers claim that many hominids died, possibly ninety percent (90%) of them. But we found, in our short distance view and individual mind, that our greatest danger is the gradual change in temperature on Earth. Nor do we consider the possibility of a meteor of significant size. NASA has ruled it out! But who is NASA? A human institution cannot guarantee the arrival of a meteorite actually more destructive and fatal that the slow process of global warming. So we need more awareness of the grains of sand that can hurt us, and we do not see that they are there. How do we prepare? Leave violence to nature, work to form an intelligent culture that allows us to communicate, and build consensus around the idea that we need to have a minimum basis for life for everyone. Those who want more can contribute more; they have the right, but everything inside a consensus of justice for all. Thus, we need a rational goal to reduce the large amount of waste we are accustomed to and redirect resources to exploring other planets, both inside and outside the solar system. Yes, violence can be a universal language, but cultivating tolerance, within the limits of justice, lets us communicate and live full/complete lives.

Then, Information is created-used in the way that each living being decides to do so. It is not dictated by any external parameter to that living being. You, before an external situation, decide to focus on solving it or escaping from it. Limitations on the use of information are also in each of us, but those limitations are reduced when working as team. Our cells take on some of our systemic information, like parts. You are responsible for another part of the information, the holistic one, you. No one has control over you; you control your mind, and with it, you should learn to measure your actions.

Chapter 8.

Levels In Informational Ability

"The evidence about Saddam having actual biological and chemical weapons, as opposed to the capability to develop them, has turned out to be wrong. I acknowledge that and accept it. I simply point out, such evidence was agreed by the whole international community, not least because Saddam had used such weapons against his own people and neighbouring countries." Tony Blair

The smallest living being that gives rise to life has minimal informational ability. Informational ability grows according to structural and functional properties of living beings, as well as other emergent properties due to the communication processes. As mentioned, the structuring of human beings is being supplemented with external tools for managing codes, which will facilitate communication and evolution towards the living being humanity, which would have the highest level of informational ability we can conceive. In the other direction, if the minimum living being is broken, it loses its informational ability. It becomes an inert matter with no informational ability. As an exercise, it is worth thinking about these days' news: we were given the news of the first artificial cell. I was surprised by the expression. Technically, they took the DNA from a cell, and replaced it with another genetic code. This cell reproduced, surprisingly, during a weekend away from the eyes of researchers. Before discussing more about the artificial cell, consider informational ability as a process.

In analyzing the fact of the artificial cell, assuming there is general truth in all this, we can say that there was an "organ transplant." The genetic code of one cell was replaced by another. The new genetic code was created in the laboratory, a great achievement. The ribosome had no problem replicating it. If we bring this situation to our human level with proper proportions, we can say that a person was handed a different model to build an artifact. The person took the device and reproduced it, using the model and the resources that were necessary and were also provided. In our structure of living matter, the nucleus of the eukaryotic cell is a prokaryotic cell. Then, an individual, the ribosome, received a model, the genetic code, with instructions to replicate it. Environmental conditions were not altered, the resources were there, and only the model was changed.

Life begins with (comes from) informational ability. Consider for a moment the first living being or group of living beings in the universe, on Earth or elsewhere; this living being has a very basic informational ability. It is now being discussed if either, a virus and another molecular structure a prion are actually living beings or not. This issue of the smallest living being is a technical matter that will be tested when we have the proper tools. Following the conceptual issue, it is known that life forms like viruses reproduce within the cells; they do not have the functionality to do so. Is the virus acting as a ribosome? Is the virus replacing the genetic code and "misleading" the ribosome? These are important technical details, but conceptually under the informational ability concept, the virus is obtaining the result of self-reproduction. Bearing this in human terms, we would have the virus like a person that has never dirtied their hands with anything, most external material functions have been performed for him, let us say a King.

Returning to the conceptual aspect of informational ability and to facilitate the explanation, go back to our structure of life. The prokaryotic cell, sometimes known as the basis for the structure of life, already has parts that are alive like the ribosome that has to know what it is doing when it creates proteins or replicates the genetic code. The ribosome is handling information because it performs quality control for genome replication. Then, at the cellular level informational ability already exists in its parts and information is used in communication. Of course, until today, everything has been explained in scientific terms (taking apart, losing the emergent properties) and with the point of view that information exists outside of living beings. Conceptually, each new level of living beings creates a new level of informational ability. Each new level of living beings creates other, more specialized functional features, creating new communication tools for that level. As one moves from one level to another, the development of more informational ability represents more capacity to discriminate, more capacity to re-present objects, which is to present once more before us, images that initially were accrued and relate them[16].

Conceptually, all humans are born with the same informational ability. When we are born, we have the ability to perceive the world as humans, to process the world in many ways, to memorize, to see colors, to hear, etc. But technically, we have different informational ability levels, our brain structure is similar but not equal, and the so called "errors" or evolution creates different DNA and different levels of informational ability, the same way daily body operations, such as eating create changes in the human beings throughout life. A material factor is the number of neurons that we are using to think, versus the number of neurons in the brain. Not all humans have the same ratio between body mass and brain mass. This relationship is often called the cephalic index. Compared to other animals, we have dramatic differences. The relationship between body mass and the mass of our brain is thirty five. We are followed

---

[16] This is the imaginative process, sub-process of informational ability, where we use curiosity and memory to play with the images and make sense of what does not exist to create, or making nonsense of the creation, to destroy.

by the chimpanzee with five point two and third, the orangutan with three, etc (Pinillos, 1970). This is one factor that tells us that we are better equipped than any other animal for processing information. Differences are not as dramatic in humans, but the point is mentioning them.

Consider the role of one being in a team, versus its informational ability with respect to other beings in the team. This item has regard to the task of the being in the team, its specialization within it. A brain cell has a different function to a cell into a muscle. The first has an informational work and the second has a material work, their functions are specialized. One processes information differently than the other, we can say they have different informational ability. Something similar happens at the level of the human being. A man and a woman have different human functions. The man contributes half the genes of the human being to be born. The woman, as well as providing the other half of the genes, nourishes, extends the fire, takes responsibility to protect it, and after birth, continue feeding it. These are quite different functions, which in some way do not influence the living beings informational process, but their particular truth. In either case, they are creating a paradigm for the protection of children and then giving a different point of reference for the informational ability; in this case, the mother generally has more affinity with that child than the father. This is conceptual and does not guarantee specific results on certain specimens.

This discussion about informational ability level from the quantity point of view opens the discussion to informational ability level from the quality point of view. There are many discussions on intelligence and consciousness, as well as on information and knowledge. Consciousness may be related to the ability to define requirements and/or set priorities. Intelligence may be related to the ability to meet needs and/or execute what is required to meet them. Information describes and defines the structures, including describing different structures with the same elements. Knowledge can define how a structure can behave in reality. We could associate intelligence with information and consciousness with informational ability.

Knowledge is unique information. Knowledge is information that is paired to facts. Knowledge cannot be transmitted because it is the direct result of living, what we experience. We use information to communicate; creating codes that are objects in our minds and become beings after the matter is transformed. As mentioned, the fish that Confucius saw was unique, and a fish that you see is unique. Similar to the concept contained in information accrual, you can have an idea of what a fish is with a drawing; you can improve the idea with a photo. But you only have knowledge when one of these animals jumps before your eyes, you feel the temperature of its body, feel the scales and body fat, etc.

As an example, let us say that a mother whose job is ironing does not want her child burned. The child has never been burned. Then, the mother communicates to the child, which does not have the experience or knowledge, "Do not touch the iron because it will burn you and it hurts." The child, out of curiosity, wants to touch the iron and has no knowledge of what it means to get burned, as we said. Here, the mother has several options, but what decision to make? Beating the child so that does not get burned? Will the child learn what it is to get burned if the mother hits the child? Lowering the temperature of the iron and letting the child learn how it feels to burned? Will she continue telling the child not to touch it? What would you do? The point in all this is the process of education. We are talking about communication and how a mother can teach her child, transmit knowledge, that the iron burns, without him getting burned. From the informational point of view, it is virtually impossible. The mother has a different pain threshold than the child; the child has less protected hands. Pain is a sensation or perception similar to listening or seeing. Over all this, what to do? Some mothers, like a mother shown in the critical periods of apartheid in

South Africa, said that she would kill her child rather than see him married to a woman of color. This is to point out the extremes to which we can go when we disagree on something, perhaps on the effectiveness of teaching methods. This could parallel our mother's decision to burn the child to teach him, or less drastically, to "let" the child, after insisting a number of times, get burned and learn on his own. Without the child touching the iron and getting burned, there is no way to communicate the sensation of pain that we feel when burned by an iron. In short, this case shows differences between information and knowledge.

We said we cannot come up with perfect ideas, but we can have solid ideas. An idea is solid if it is near the general truth. We have solid ideas when we review the system and the process from several points of view and give various interpretations, conducting an assessment that allows consistency between the object and/or the model and what the objective or purpose of the idea is. That is, if we have the goal in mind, it is solid if the system is consistent, its properties and limits are indistinctly defined, and the process that is pursued, once it is reviewed from various points of view, generates different interpretations, where at least most of these interpretations are coherent. We are aligning our informational ability to obtain the best information possible i.e. obtain proper meaning from what is perceived.

Intelligence And Consciousness

Intelligence is a process related to specific issues about perceptions, which we call actions. Intelligence can be understood as the ability to discern and follow rules when discriminating actions. It is born from statistical processes and allows us to analyze how systems are structured and how processes function.

Consciousness is a process related to abstract issues about discriminations, which we call models (objects). Consciousness can be understood as the ability to construct and make sense of elements, where there is no pre-establish rules and creating what does not exist synthesizing.

Some living beings create problems; others resolve them. We are talking about our brains. One part of the brain can solve one transportation problem. The other part may create a problem, finding that the transport solution is not so pleasant.

Consciousness is part of our informational ability. From the humankind level we can relate it to morality, an abstract model of behavior. Let us look at a moral classification by a specialized book. James Rachels –Stuart Rachels in *The Elements of Moral Philosophy*, speaks of the states in Kohlberg's moral development. Starting at the basic level:

- Obeying authority and avoiding punishment (stage 1).
- Satisfying one's own desires and letting others do the same, through fair exchanges (stage 2).
- Cultivating one's relationships and performing the duties of one's social roles (stage 3).
- Obeying the law and maintaining the welfare of the group (stage 4).
- Upholding the basic rights and values of one's society (stage 5).
- Abiding by abstract, universal moral principles (stage 6).

Look at what Mira y Lopez says about duty, which because of the abstract notion may be related to the highest moral concept, as well as other abstractions, such as fear, anger and love, which ultimately have nothing to do with intelligence, but with the consciousness of every individual. Usually, the answer people give to a question on one judgment of consciousness is "I do not know." Example: Why do you love him? I do not know. It is because... Why does it enrage you? I do not know. It is because... etc. Let us now see what Mira y Lopez wrote:

"Well then, there are three primary emotions in which the full range of reflections and deflections of escape, assault and possessive fusion are inscribed. Their most common names are: FEAR, ANGER and affection or LOVE. The energy which they are capable of mobilizing and vehicular is so immense, that whatever mankind has done, of good and of evil on Earth, must be fundamentally to its account. But, for many centuries, human beings have not lived isolated and anarchically on the planet's crust, but they constitute groups, and therefore, each individual requires – by willingness or by force – the category of 'homo socialis'. And here comes in another huge force, predominantly repressive of those former, which is commonly known by the name of law, duty, customs, norms, traditions, etc. not only contained in codes and commandments more or less sacred, but stored in certain 'authorities', who use their power to ensure that it is introduced equally in each brain, right when is able to receive it. We are going to call that fourth force overall, DUTY.

Certainly, it is not possible to consider this new face in the same plane as those above; it is not, in the first place, congenital, neither can it be included in the generic qualifier of emotions. But, as we shall see at the appropriate time, it is capable, many times, to shock the man and make him, sometimes, resist the onslaught of any of them, or even, all of them together. Like fear, anger or love, duty, when it is not satisfied, can not only bite, but also prick in the mood bowels and lead to the greatest suffering and to suicide. It can, thus, compare, without lessening, with the three 'natural' giants, this 'social' giant which, in a sense, is derived from them, and contains something from each one in its unique texture."

Intelligence and consciousness can be seen then as reason and emotion. We can see intelligence as the study of physics, the analytical process, the study of parts and their interactions, separating wholes. We can see consciousness, as the study of systems, the synthetic process, the study of wholes and their exteractions, integrating parts. The former, intelligence, discriminates and compares objects, studies possible interactions and discovers their operation, identifies the function(s). The later, consciousness, as an informational process, also discriminates and compares objects, but unlike intelligence, these objects are synthesized under abstract classifications, creating the models already mentioned, coming to discover purely informational connections, such as intention(s) and objective(s).

To bring these concepts to more familiar ideas, let's say that intelligence is given by IQ, the Intelligence Quotient or logical intelligence, reasoning. Consciousness is given by EI, Emotional Intelligence, a term coined by Daniel Goleman of which I still do not know an index, but could be developed. Bringing these to the traditional model of the brain, we would have one lobe of the brain having reasons and the other lobe having emotions, all at the same time. Of course, if you are cold, purely material, tilted only to reason, let us say insensitive, you will think differently than if you are warm, totally ideal, leaning only to emotion, let us say sensitive. Here, we can talk about parallel ideas. One is not related to the other; they don't touch but are complementary. Culturally we have given more importance to reason, logic. But emotions mark more the road of powerful people, some of them ambitious idealists, with a large optimism and with an impressive practicality, where the end justifies the means. These are usually confidence men, who see managing people in a social environment like managing automatic machines, solving problems for some logical people. The Benziger model is perhaps more appropriate in talking about concreteness and abstraction. The basal parts of both lobes handle the concretion, intelligence, and the front parts of both lobes handle abstraction, consciousness.

In theory, one can say that one has only one of these two classic members of the mind, intelligence or consciousness, but not so in practice. In practice, there are levels, no one is completely rational, and no one is completely emotional. The different levels of rationality and sentimentality create countless particular truths, a countless number of informational abilities. Intelligence allows us to physically take care of ourselves, thinking of optimizing processes beyond people. We need to think of ourselves. Consciousness allows us to take care of others, emotionally, caring for those who are "close" to us. We require thinking of others' needs. Here, intelligence and consciousness are disequilibrium to control, one on which we continuously work on: Who is worth more, you or me? Who is worth more, a woman or a man? Who is worth more, an adult or a child? Some of us are taught to think first about others, but is that right? Others were told to think first about themselves. Is that right? Everyone has to decide with his own criteria. We all have reasons and emotions on which to think before acting, reasons to appear as good to others, or emotions, to feel that we should be good and that we can have a better world.

From another point of view, informational ability can be seen as effectiveness. Intelligence talks about efficiency and efficacy. Consciousness talks about ethics and elegance, as seen in Checkland's 5 E's model (Checkland). Effectiveness is given by the right balance between intelligence and consciousness. We believe in being efficient and efficacious with inert beings and to respect the ethics and elegance of other people and ask to have ours respected. Intelligence leads us to improve equipment and reduce costs and waste. Consciousness leads us to respect others and to earn respect. When it comes to respecting, we set the level of respect we give. When it comes to being respected, we decide on the level of respect we receive. In either case, our particular truth is the measure of respect, talking about self-esteem or respect for others. Eleanor Roosevelt said that we are only disrespected when we accept what is said to us. She said it from the informational point of view.

For human beings, informational ability has a different order or level. We can say it from experience, seeing the way human beings have asymbiotically changed nature. But we do not have the exclusive domain of consciousness, as mentioned by some authors. Other living beings transform nature, but they do so from a lower level, from a lower level of abstraction than us humans. Humans, with the development of complex codes such as writing, allowed by our high abstraction ability, have created external memory for the human being. These external memories really set us apart from any other animals. As mentioned, we started in caves, with simple lines on the walls, then on papyrus, followed by hieroglyphs, writing on paper, and now programming computers; all these tools have enhanced our information ability. Without them, we would not be where we are technologically, communicating from one human to another human without the former being close to the later. So ideas have spread and been supplemented, creating synthesizer and quantifier languages like mathematics, which has greatly helped to extend informational ability and create new knowledge.

Solid Ideas, Information Through All Distance

We have said that life is an integral issue between matter and information. Your particular truth contains all the information you have accrued by any means, inheritance or experience. The conventional truth is the information we share with others, according to each group. The general truth is matter with all its structures and emergent properties.

Ideas could be considered solid if they pass the three tests of truth: the particular truth, the conventional truth, and the general truth.

To create an idea, the living being must follow an informational process. To create means to materialize the idea, perform material transformations. We can say that matter interacts and is self-perceived; the interactions are changes by living and non-living beings, some which we perceive. Our formula for life summarizes the process of creation by the living being; to live = (P+E)/C. Our living being creations are accrued as information within us, and with the help of matter, the end stage, the results are there for the final judgment: the idea is created –the matter is transformed.

The first stage in testing the idea is the proof of the particular truth. You, as the individual that generates it with your informational ability, using your intelligence and consciousness, calibrate the idea. Of course, when you are thinking, there is an interchange of information between cells or groups of them within you, which occurs by material means. In this exchange, you are the reference for the process, a very "high" reference because you are a system of approximately 10^{28} particles. But because in these cycles you are the reference, you are a whole creating ideas. At some point in the creation of the idea, you decide that the idea is mature; you already validate in this first stage. Who decided that the idea was sound? You did. Who will win if the idea is as solid as you think? You will. Your team of 10^{28} particles that structure you, give the verdict, the idea is solid.

The second stage in testing the idea is the proof of the conventional truth. At this stage, you, as the author of the idea, are starting an informational action out of you, an exteraction. You are sharing information, communicating with another human. The smallest system that exists is formed by two units. This step includes you and another person. In the first step of this second stage, you discuss your ideas with the other person. After communicating with that person, that person will make a final judgment, where he has used his particular truth and informational ability on his own. The final judgment of the other person is whether the idea is solid or not. If passed, the idea has taken a step towards passing the conventional truth. But one swallow does not make a summer; it requires a critical mass, even in the presence of a catalyst. You will work step by step to obtain the critical mass, so your idea is validated by stakeholders and passes the test of conventional truth. By having the right proportions, the idea is considered valid, it has conventional truth validation. As we have said regarding perfect decisions, there are many elements involved with one idea, which are informational and material. Each living being does not have the capacity to realize all points of view, in all densities and all interactions. Then, teams are essential in the process of testing the solid ideas' second step.

The third stage in testing the idea is validating it toward the general truth. At this stage, you, as the author of the idea and other "required" individuals, as co-authors, who validated the idea in the second stage, work together by exchanging information, using resources and tools to transform matter and materialize the idea, to construct what was defined as solid idea. You can only check the soundness of an idea when you pass this last stage, not before. This concept of robustness can be applied to any idea. We are not talking about perfection; we are talking about consistency; what the idea was said to be, or wanted to produce was created and can be observed when comparing the specifications given to the product or the service at the beginning of the project, with what comes at the end of the project. In other words, the product or the service reflects what was designed.

Let's look at an example, a couple's marriage. In this case, the idea of marrying comes out from one of the two people. The author of the idea has negotiated with himself and reflected according to cultural parameters and the knowledge of the counterparty, whether it is or it is not the time. If the analysis is correct, the other party should accept immediately. According to Benziger, usually the couple's counterpart complements his informational ability. Quite possibly, a negotiation process starts. In the process there are conditions according to culture: "We have no home," "We have not decided how many children we will have," or "We do not know ourselves well enough to take that step," etc. Finally, after the negotiation process and some material movements, the couple will communicate the decision to other parties. Let us say, at this point the idea will have passed the conventional test. The idea will be judged solid at the end of the marriage, depending primarily on the reason for separation. Let's say that if after a "reasonable" period of years one of the two dies, then the idea can be judged as solid if they had an independent house, children, and were "happy." You can take a project to judge the soundness of the concepts and define whether the idea is solid or not.

The levels in informational ability can be identified when looking at the structure. One prokaryotic cell has less informational ability than one eukaryotic cell and these, less informational ability than multicellular colonies or living beings like us. The "big" difference between us and other animals is not that we possess a unique informational ability; it is the level or degree of informational ability. Now, when we look at the informational processes, consciousness and intelligence, we can understand that any living being has some degree of consciousness and some degree of intelligence. Consciousness identifies, defines situations, assigns priorities, and is strategic; intelligence uses restrictions, manages resources, maintains discipline, and is tactical. If there is no basic goal, the desire to live, which is given by consciousness, there is nothing to resolve on the part of intelligence.

Summarizing the concept of life, we cannot see ourselves as a block of matter with life; we are complex structures of living beings, organism of organisms. These organismic structures start with a minimum of informational ability and come to different capacities depending on the structure. With more knowledge about the informational process, we will find where the division of living being loses its informational ability. There is so much instinct in humans as in other animals, but the ability for abstraction on us humans lets us build unique informational concepts, with which we have distanced ourselves from. Integrating the concepts of use of information and levels of informational ability, we can understand that humankind is a living being. The use of information technology will assure over time, as in the growth process of every living being, that humanity functions as the living being that it is and like an organism, involves all its parts in the process, and when that time comes, the superman will exist:

> *"And the superman will not renounce its animal origin, nor will it continue determined to appear to be what it is not: but will know how to reduce the gap between its ideal dreams and its practical achievements. That day, from words to actions, there will not be a long way; from what was promised to what was fulfilled, will just be an axiological distance, the coercion of the group will be imperceptible; the competition there will be transformed into cooperation, there will be no struggle for life, but emulation in life; there will be no aspiration, to be rich, nor famous, nor powerful, nor genius, nor holy... will aim at making one's life, one harmonious play – that by doing so will be fair and beautiful – in relation to the great universal life, which we are, just instantaneous and tiny moments."*
> (López, 1965).

For now, we must understand that every living being creates its own life with faith in its future plans and with the courage to fight for them.

Bibliography

Alonso, M., & Finn, E. J. (1992). *Physics*. London: Addison Westley Longman Ltd.

Arango, C. (2011). *Comments By Catalina*. Miami: Catalina Arango.

Arango, J. D. (2010, 7 21). *Information and living things*. Retrieved 9 23, 2010, from ISSS - International Society of Systems Science: http://journals.isss.org/index.php/proceedings54th/article/view/1391

Ashby, W. R. (1957). *An Introduction To Cybernetics*. London: Chapman & Hall Ltd.

Bertalanffy, L. V. (1976). *Teoria General de los Sistemas* (1ra edicion ed.). (J. Almela, Trans.) Bogota: Fondo de cultura economica.

Checkland, P. B. *Soft Systems Methodology in Action*. John Wiley & Sons.

Colvin, G. (2008). *Talent Is Overrated: What Really Separates World-Class Performers from Everybody Else*. New York: Penguin Group.

Covey, S. R. (1990). *The 7 Habits of highly effective people*. New York: Fireside.

Dupage. (n.d.). *Prokaryotic and Eukaryotic Cells*. Retrieved 09 29, 2010, from College of Dupage: http://www.cod.edu/PEOPLE/FACULTY/FANCHER/ProkEuk.htm

Jacob, E. B., & Shapira, Y. (2004). *Meaning-Based Natural Intelligence...* Tel Aviv: Tel Aviv University.

Katherine Benziger, P. (2006). *Thriving In Mind*. Carbondale: KBA, LLC.

Language., T. A. (n.d.). *Information*. Retrieved 12 8, 2010, from The Free Dictionary: http://www.thefreedictionary.com/information

Larousse. (2007). *Diccionario Manual de la Lengua Española*. Vox: Larousse Editorial, S.L.

Lengua, R. A. (2009). *WordReference.com From the Real Academia Española - Vigesima segunda edicion*. Retrieved 09 24, 2010, from Word Reference: http://www.wordreference.com/es/en/frames.asp?es=vida

Livas, J. (2009, 07 06). *Stafford Beer*. Retrieved 10 07, 2010, from Youtube: http://www.youtube.com/watch?v=7O09FPHuCQQ

López, E. M. (1965). *Cuatro gigantes del Alma*. Buenos Aires: El Ateneo.

Periodic Table of Elements. (n.d.). Retrieved 09 28, 2010, from Periodic Table of Elements: http://www.ptable.com/

Peter, L. J., & Hull, R. (1979). *El Principio de Peter*. Barcelona: Plaza y Janes S.A.

Pinillos, J. L. (1970). *La Mente Humana*. Navarra: Salvat Editores S.A.

Rodriguez G., A. (2003). *Artefactos, Diseño Conceputal*. Medellín: Fondo Editorial Universidad Eafit.

Ropke, W. (1957). *Más allá de la oferta y la demanda*. Ginebra: Ropke.

Rosen, R. (2003). *Anticipatory Systems*. Rosen Enterprises.

Russell, B. (1949). *Autoridad e Individuo*. México: Fondo de Cultura Economica.

Utah, U. o. (n.d.). *Cell Size and Scale*. Retrieved 12 2, 2010, from Learn.genetics: http://learn.genetics.utah.edu/content/begin/cells/scale/

Wikipedia, v. e. (n.d.). *Historia de la astronomia*. Retrieved June 30, 2010, from http://es.wikipedia.org: http://es.wikipedia.org/wiki/Historia_de_la_astronom%C3%ADa

Glossary:

1-2-3 or one, two, three: 1 being, 2 objects, 3 elements. Mnemonic tip to remember that when we (you and me) are in front of *one* being, there are *two* objects, one in your mind other in my mind, and overall, there are *three* elements; one being plus two objects. Let us have an example. There is one glass. The representation of one glass by the reader creates one object; a writer's representation creates another object; then we have two objects. An element is any being or object and in this case, we have three elements. The being –the glass– is one element, the two objects, yours and mine, are two elements, and combining all, 1 being plus 2 objects equals 3 elements.

Abstraction: See Informational Ability.

Action: Disruption of a particle or group of them. An action is the most basic step in a process. One action changes the speed (direction) of a moving particle or a group of particles or alters its charge distribution. An action, when discriminated by living beings, creates an event.

Atome: An indivisible particle. This word renames the word atom. This is done because the meaning of the word has been lost. Atom, is used to refer to a group of particles.

Being: Any particle, or group of particles. It is a material density; beings have mass and charge. A being is tangible. It is something that exists in itself, does not require informational ability to exist.

Control: Results of combined actions to direct beings toward an objective. Control requires attribution/creation of meaning.

Control derives from the concept of dynamic equilibrium, where a being forces another to maintain disequilibrium toward a target or objective. The force of gravity, combined with the material inertia, creates the idea of control among celestial bodies. We can say that the Earth "controls" the orbit of the moon, but this is one balance, mutual action between the gravity of both celestial bodies and the inertia of the mass of the moon, that tries to keep the moon in a linear motion.

Conventional Truth: All the elements that contribute to form the culture of one community or group of living beings, which the group believes is valid. The conventional truth can be seen like the informational limits of the community. The conventional truth is unique to each group, regardless of its size. There are opportunities/threats (crisis) in a community because there is difference in the conventional truth. The information shared is conventional for the group; the one which is not shared belongs to each individual's particular truth or sub-group's conventional truth. We can say that the conventional truth get its way by sharing elements of particular truth.

Discriminate: Delimiting a perception to create initial meaning. Discrimination is a sub-process of informational ability, which assigns limits or creates references from perceptions. An example is an abstract painting; when seeing it, there are many decisions about what it represents. When we discriminate, we create "raw" objects.

Disequilibrium: The "flow" of elemental particles. Memory is a prerequisite in order to discriminate disequilibrium, since the previous or next position of the elemental particle do not exist at the present.

Element: Basic informational notion or its material counterpart. An element can represent one being, one object, or an action between them. Thus, an element can be a name for anything.

Emergent: That arises "spontaneously". Emergent refers to a new structural disequilibrium that arises by combined actions among elemental particles or groups of them. On the information side, there are emergent ideas when thinking, since we move ideas (objects) in our mind compare them and have an emergent new idea. Reviewing the definition of control, we find that celestial bodies Earth-moon arrive "spontaneously" to their balance. In this case, there is a process where there is not the number of actions before the emergent property, but the one that has success, that is what spontaneously refers to.

Event: Discrimination of a step in a process. In order to exist, an event needs to be discriminated. An event is an informational concept between two states of matter. An event is composed of one or many actions, among the limits defined by the discrimination. One event can be viewed as the transition between two states of matter. One gunshot is an event, a process. This event has several sub-processes: pulling the trigger, hitting the primer, output of the missile, cap exit etc. which in turn are events, because there are sub-process, actions between these two states of the matter.

Execute: Transforming the structure of one part or piece of matter, having a plan. Execution is an event performed by (a) living being(s).

Exteraction: It is an action among systems. An exteraction is, conceptually, an action between systems, one system to another system or its environment. In reality one system actions can disequilibrate others with a single exteraction. The universe, by being all that exists, has no exteractions. By executing actions, the living being is part of the universal process, where the living being represents a system and executes external actions called exteractions.

Fact: Group of particles or their interactions. A fact represents the change in the relative position of elemental particles, or a specific structure of elemental particles.

Information: The result of informational ability. Information is attribution/creation of meaning. Information represents the nature for the living being that has done the informational process. In some cases, it is difficult to distinguish if the result, information, is either: one being or one object, one system or one process, one end or one mean.

Informational ability: A process where information is created. The essential process of living beings, its result is meaning production. This process is the essence of life. Informational ability, as creation/attribution of meaning, is only possible for living beings.

Interaction: Action among particles or among parts of a system. Conceptually, it is an action within a system, between two parts. In reality, one internal action –from one part– can affect many other parts. Basic interaction is an action between particles. If all that exists are particles, which we call matter, there are only interactions. In this case, all actions occur internally within the universe. Consider interactions without purpose until the emergence of informational ability that still depends on material activity, but starts giving direction to the material activity.

Life: A synthetical material property that emerges among material structures and their disequilibrium. This property lets this matter direct matter. The being that has life is called living being. The essence of living beings is the capacity to choose direction which comes from a process, informational ability. Life is a controlled process with the objective to live. Sustaining life requires the use of information. Life requires structure, disequilibrium, and information to direct matter.

Limit: Informational concept that allows living beings to create objects. Without applying limits, we cannot discriminate. Every living being defines limits in order to achieve discrimination.

Matter: The group of all elemental particles. Elemental particles have two properties, mass and charge. Matter is responsible for all that exists, since through its structure and disequilibrium, it is able to synthesize life. The set of all the particles that exist and their interactions make up the universe. The minimum "natural" particle is the atome.

Model: Or second order objects. Models are more abstract objects, frequently with no counterpart in reality but existing in nature. Examples are: colors, sounds, fear, love and many others. Using abstraction, we remove elements from objects for comparison. Another example: Geometry is built based on models. The Euclidean point does not exist, since if it had any measure at all, it would be a sphere.

Nature: The result of structural matter disequilibrium from life's point of view i.e. when information is created. The model for nature presents life as the results of direct matter in an informational process with at least four steps, which are: birth, growth, reproduction, and death.

Object: The result of the informational ability process when it is applied to structures and disequilibrium. The representation of a being or the distinction when beings are compared. It is the result of an action by the informational ability of one living being. An object is intangible. It represents a being; it is made by one living being. Objects need to be created by living beings. By abstraction level there are first and second order objects. First order objects can be called facts. Second order objects can be called models.

Objective: Set of limits that define a process or system to be reach.

Part: Group of elemental particles that easily follows a structural pattern within a system. We are familiar with body parts. Example: eyes, ears, heart, etc. are parts of a body. Here the pattern is conventional.

Particular Truth: All the elements that contribute to forming one living being philosophy; what one living being believes is true; its true nature. The particular truth can be seen like the informational limits of living beings, or the reality of each individual. The reality of each of us is based on the information we have from ourselves and the outcome of our actions with our environment. The particular truth changes as we live. The particular truth is unique to each individual.

Perception: The event of receiving an action. Perception is taken directly from elemental particles shocks, or indirectly generated by shock's waves that create disequilibrium.

Piece: Group of elemental particles that does not follow a structural pattern within a system. It is a fraction of one system or from one part of it. If your body is cut in two halves, many of your organs are going to lose the structural pattern, the form they have in your body.

Planning: Thinking with action in mind. (Dictionary)

Point of view: Standpoint a living being takes on a fact. There are material and informational points of view. The material point of view reflects the position you are using –the front, back, lateral, etc. The informational point of view reflects the paradigms and other beliefs you have and are used by your informational ability. Your points of view come from your particular truth and your informational ability.

Problem: A mismatch between expectations and results. Any problem is a matter of information.

Process: The set of actions coming from matter disequilibrium. The informational ability comes from the emergent properties of a material process. Processes have two basic classes: transformation and transport. Transport can be identified as a transformation of a system, because changing the position of one or more of its elements creates a new structure. A new structure represents a transformation of the former.

Reality: The result of matter structural disequilibrium. It is a model that presents elemental actions of matter creating and recreating structures.

State: A state is given by the idea of freezing a set of elemental particles at a given time. One state reflects the place of elements, objects, or beings; defining a frozen system. Without informational ability, it is not possible to derive the change in the actions of matter, those actions that exist between states. In common concepts, a state is equal to a photograph.

Step: Part of one process –sub-process. One step can have multiple actions. The steps are defined as events by the living being according to its informational ability.

Structure: Is the "place" of elemental particles at a given time. Criterion is a prerequisite in order to discriminate structure, since limits do not exist in reality. The interaction among the particles allows particles that are forming groups, which we call beings, to acquire emergent properties that meet that specific combination of particles (structure). The same group of particles in different combinations (structures) creates different groups with different attributes (emergent properties).

System: The set of limits coming from matter structure. A system is a group of particles. To create a system requires informational ability. A system is a "sub-universe" and what is excluded is its environment. A living being is considered a system. Systems are structure in two basic concepts: components and links. Components are the limits where transformation takes place. Links are the conduits or supportive elements that let transportation take place.

Thinking: The act of using informational ability. An action of a living being that creates new abstractions-discriminations.